FOREWORD

It is a common fact that students do not show much interest in solving problems in Integral Calculus when compared to that of Differential Calculus. The voluminous nature of the problems in Integral Calculus forbids the students to gain confidence in this subject.

Have a look on the following discussion. A question was asked by a student and was explained by an user in the internet.

Question: *I have never done integration in my life and I am in the first year of university. Is it (integration) harder than taking the derivative? I've heard it just going backwards. Is it generally considered harder than differentiation?*

Explanation given: If you are fine with derivatives, you will be fine with integrals in I year calculus. It never hurts to pay attention in class and to do your homework[1]. .... In fact, if you have trouble with a problem, you should do more of the same kind as soon as you know the answer[2]. .... The kind of problems you get in first year calculus will be solvable if you learn enough tricks[3]. .... Integrals start out harder than derivatives and wind up easier[4]...........

Superscript 1 means '**Be familiar with the formulae and methods of solving problems in Differential Calculus and Trigonometry**'.

The formulae practice workbooks in Differential Calculus and Trigonometry (**PROF MSDOSS MATH BOOK SERIES I and II**) help the students to achieve this.

Superscript 2 emphasize on '**Practice! Practice!**'

Students gain confidence only through practice only. This can be achieved by following the methods explained in the formulae practice workbooks in Differential Calculus, Trigonometry and Integral Calculus (**PROF MSDOSS MATH BOOK SERIES I, II and III**)

Superscript 3: 'Trick' means '**Ability to understand and classify the problems!**'

The above trick is rightly followed in the above mentioned formulae practice workbooks.

Superscript 4 indicates the **outcome!**

Experience shows that the above mentioned workbooks help the students to achieve this result.

Significant features :

- Each unit is provided with a revision of the formulae applied and methods followed.

- Self- evaluation test is provided at the end of each unit.

- Already tested in India among the average and below average students with good results.

- Definite integrals, evaluation of integrals using partial fraction and the remaining methods of evaluation of integrals will be discussed in volume II.

Prof. M. SUBBIAH DOSS

PROF. MSDOSS MATHS BOOK SERIES III

CALCULUS II

INTEGRAL CALCULUS

Formulae Practice Workbook

By Prof. M. Subbiah Doss ©

Author's email- id : subdoss2014@gmail.com

First Edition : December, 2016

Pages : 83

CONTENTS

Calculus II – Integral Calculus workbook - Vol I

INTEGRATION

There are two methods of integrating a given function in Integral Calculus.

1. Integration as the inverse process of differentiation

2. Integration as the limit of a sum of a certain series.

In this workbook, we follow the first method only.

1. Integration as the inverse process of differentiation

Different methods are available here. One must learn to select and apply the appropriate
method to integrate the given function. A regular practice is required for this.

As in the case of differentiation, integration also satisfies the property of linearity.

ie. $\int k f(x)\, dx = \int f(x)\, dx$ and

$\int (f_1(x) \pm f_2(x))dx = \int f_1(x)\, dx \pm \int f_2(x)\, dx$

where k is a constant.

UNIT 1 Integration as the inverse process of differentiation

Integrals of the form

UNIT 1.1 $\int x^n\, dx, \int \frac{1}{x} dx, \int (ax + b)^n\, dx, \int \frac{1}{ax+b} dx$

UNIT 1.2 $\int e^x\, dx, \int a^x\, dx$ and $\int e^{ax+b}\, dx$

UNIT 1.3 $\int \sin x\, dx, \int \cos x\, dx, \int \sin(ax + b)\, dx, \int \cos(ax + b)\, dx$

UNIT 1.4 $\int \sec^2 x\, dx, \int \operatorname{cosec}^2 x\, dx,$

$\int \sec^2(ax + b)\, dx, \int \operatorname{cosec}^2(ax + b)\, dx$

UNIT 1.5 $\int \sec x \tan x\, dx, \int \operatorname{cosec} x \cot x\, dx, \int \sin x \cos x\, dx$

$\int \sec(ax + b)\tan(ax + b)dx, \int \operatorname{cosec}(ax + b)\cot(ax + b)dx$

UNIT 1.6 $\int \frac{1}{\sqrt{1-x^2}} dx, \int \frac{1}{1+x^2} dx, \int \frac{1}{1+(ax)^2} dx,$

$\int \frac{1}{\sqrt{1-(ax)^2}} dx$ ....

1.1 Integrals of the form $\int x^n \, dx$, $\int \frac{1}{x} dx$, $\int (ax+b)^n \, dx$ and $\int \frac{1}{ax+b} dx$

Let us differentiate $\dfrac{x^{n+1}}{n+1}$ w.r.to x.

$$\frac{d}{dx}\left(\frac{x^{n+1}}{n+1}\right) = \frac{1}{n+1}\frac{d}{dx}(x^{n+1}) = \frac{1}{n+1}(n+1)x^n = x^n$$

By differentiating $\dfrac{x^{n+1}}{n+1}$ with respect to x, we get x^n.

As integration is the inverse process of differentiation,

by integrating x^n we must get $\dfrac{x^{n+1}}{n+1}$. Hence

$$\int x^n \, dx = \frac{x^{n+1}}{n+1} + c$$

Exemption :   Here $n \neq -1$. Why? If $n = -1$, what happens?

$$\int x^{-1} \, dx = \frac{x^{-1+1}}{-1+1} = \frac{x^0}{0} = \frac{1}{0} \text{ an indeterminate form.}$$

Hence the correct formula is

$$\int x^n \, dx = \frac{x^{n+1}}{n+1} + c \, , \, n \neq -1$$

When $n = -1$, we have a separate formula $\int x^{-1} \, dx = \int \frac{1}{x} dx = \log x$

[Reason : $\dfrac{d}{dx}(\log x) = \dfrac{1}{x} ==> \int \dfrac{1}{x} dx = \log x$]

$$\int x^{-1} \, dx = \int \frac{1}{x} dx = \log x + c$$

In particular, $\int dx = \int x^0 \, dx = \dfrac{x^{0+1}}{0+1} = x$. Thus $\int dx = x + c$

Examples 1.1.1

1. $\int x^5 \, dx = \dfrac{x^{5+1}}{5+1} = \dfrac{x^6}{6} + c$

2. $\int \dfrac{1}{x^7} dx = \int x^{-7} \, dx = \dfrac{x^{-7+1}}{-7+1} = \dfrac{x^{-6}}{-6} + c = -\dfrac{1}{6x^6} + c$

3. $\int \sqrt{x}\, dx = \int x^{\frac{1}{2}}\, dx \quad = \dfrac{x^{\frac{1}{2}+1}}{\frac{1}{2}+1} = \dfrac{x^{\frac{3}{2}}}{\frac{3}{2}} + c = \dfrac{2x^{\frac{3}{2}}}{3} + c$

4. $\int \sqrt[3]{x}\, dx = \int x^{\frac{1}{3}}\, dx = \dfrac{x^{\frac{1}{3}+1}}{\frac{1}{3}+1} = \dfrac{x^{\frac{4}{3}}}{\frac{4}{3}} + c = \dfrac{3x^{\frac{4}{3}}}{4} + c$

5. $\int (1 - 3x^2)\, dx = \int dx - 3\int x^2\, dx = x - 3\dfrac{x^3}{3} + c = x - x^3 + c$

General form (integrand is a linear function of x)

Since $\quad \dfrac{d}{dx}\left(\dfrac{1}{a}\dfrac{(ax+b)^{n+1}}{n+1}\right) = \dfrac{1}{a(n+1)}\dfrac{d}{dx}\left((ax+b)^{n+1}\right)$

$$= \dfrac{(n+1)(ax+b)^n.\, a}{a(n+1)} = (ax+b)^n,$$

$$\Longrightarrow \int (ax+b)^n\, dx = \dfrac{1}{a}.\dfrac{(ax+b)^{n+1}}{n+1} + c$$

Also we have, $\quad \dfrac{d}{dx}\left(\dfrac{1}{a}\log(ax+b)\right) = \dfrac{1}{a}\dfrac{d}{dx}(\log(ax+b))$

$$= \dfrac{a}{a(ax+b)} = \dfrac{1}{ax+b}$$

$$\Longrightarrow \int \dfrac{1}{ax+b}\, dx = \dfrac{1}{a}\log(ax+b) + c$$

Thus, we have the general formulae

$$\boxed{\begin{aligned} \int (ax+b)^n\, dx &= \dfrac{1}{a}.\dfrac{(ax+b)^{n+1}}{n+1} + c \\[2mm] \int \dfrac{1}{ax+b}\, dx &= \dfrac{1}{a}\log(ax+b) + c \end{aligned}}$$

Examples 1.1.2

1. $\int (3x+5)^5\, dx = \dfrac{1}{3}\dfrac{(3x+5)^6}{6} + c = \dfrac{(3x+5)^6}{18} + c \qquad\qquad (\,a = 3, n = 5\,)$

2. $\int \dfrac{1}{(4x-3)^6}\, dx = \int (4x-3)^{-6}\, dx = \dfrac{1}{4}\dfrac{(4x-3)^{-5}}{-5} + c = \dfrac{(4x-3)^{-5}}{-20} + c\ (\,a = 4,\ n = -6)$

3. $\int \sqrt{px+q}\, dx = \int (px+q)^{\frac{1}{2}}\, dx = \dfrac{(px+q)^{\frac{3}{2}}}{\frac{3}{2}p} + c = \dfrac{2\,(px+q)^{\frac{3}{2}}}{3\,p} + c\ \ (\,a = p,\ n = \frac{1}{2})$

---

4. $\int \sqrt[5]{7x+3}\, dx = \int (7x+3)^{\frac{1}{5}}\, dx = \dfrac{(7x+3)^{\frac{6}{5}}}{7 \cdot \frac{6}{5}} + c = \dfrac{5\,(7x+3)^{\frac{6}{5}}}{42} + c$ ($a = 7$, $n = \frac{1}{5}$)

5. $\int \dfrac{1}{\sqrt{(2x-3)}}\, dx = \int (2x-3)^{-\frac{1}{2}}\, dx = \dfrac{1}{2}\dfrac{(2x-3)^{\frac{1}{2}}}{\frac{1}{2}} + c = \sqrt{(2x-3)} + c$ ($a = 2$, $n = -\frac{1}{2}$)

Note : In the above example, by integrating $\dfrac{1}{\sqrt{(2x-3)}}$ we get $\sqrt{(2x-3)}$. The other way must be true. ie. by differentiating $\sqrt{(2x-3)}$, we must get $\dfrac{1}{\sqrt{(2x-3)}}$. Isn't it?

Let us verify this :

$\dfrac{d}{dx}(\sqrt{(2x-3)}) = \dfrac{1}{2\sqrt{(2x-3)}} \cdot 2 = \dfrac{1}{\sqrt{(2x-3)}}$. o.k!

Similarly we can verify this for the remaining problems.

6. $\int \dfrac{1}{8x+1}\, dx = \dfrac{1}{8}\log(8x+1) + c$ ($a = 8$, $b = 1$)

7. $\int \dfrac{1}{9x-2}\, dx = \dfrac{1}{9}\log(9x-2) + c$ ($a = 9$, $b = -2$)

Remember the following

$\int x^n\, dx = \dfrac{x^{n+1}}{n+1} + c\,, n \ne -1.$

$\int x^{-1}\, dx = \int \dfrac{1}{x}\, dx = \log x + c$; $\int dx = x + c$

$\int (ax+b)^n\, dx = \dfrac{1}{a}\cdot\dfrac{(ax+b)^{n+1}}{n+1} + c$; $\int \dfrac{1}{ax+b}\, dx = \dfrac{1}{a}\log(ax+b) + c$

Exercise 1.1

1. $\int \dfrac{1}{\sqrt{x}}\, dx$ = ...

2. $\int \dfrac{1}{(3x-2)^{\frac{3}{2}}}\, dx$ = ...

3. $\int \dfrac{1}{2x+7}\, dx$ = ...

4. $\int (2-9x)^{12}\, dx$ = ...

5. $\int \sqrt[3]{5x+2}\, dx$ = ...

6. $\int \dfrac{1}{4x}\, dx$ = ...

7. $\int (5+x)^{11}\, dx$ = ...

8. $\int \dfrac{1}{x+1}\, dx$ = ...

9. $\int \dfrac{1}{5-2x}\, dx$ = ...

10. $\int \dfrac{1}{(3x+4)^5}\, dx$ = ...

Solution 1.1

1. $\int \frac{1}{\sqrt{x}} dx = \int x^{\frac{-1}{2}} dx = \frac{x^{\frac{1}{2}}}{\frac{1}{2}} + c = 2x^{\frac{1}{2}} + c$

2. $\int \frac{1}{(3x-2)^{\frac{3}{2}}} dx \quad = \int (3x-2)^{\frac{-3}{2}} dx = \frac{1}{3} \frac{(3x-2)^{\frac{-1}{2}}}{\frac{-1}{2}} + c = -\frac{2}{3(3x-2)^{\frac{1}{2}}} + c$

3. $\int \frac{1}{2x+7} dx \quad = \frac{1}{2} \log (2x + 7) + c$

4. $\int (2 - 9x)^{12} dx \quad = \frac{1}{-9} \frac{(2-9x)^{13}}{13} + c = \frac{(2-9x)^{13}}{-117} + c \quad (a = -9, b = 2)$

5. $\int \sqrt[3]{5x + 2} \, dx \quad = \int (5x + 2)^{\frac{1}{3}} dx = \frac{(5x+2)^{\frac{4}{3}}}{5.\frac{4}{3}} + c = \frac{3(5x+2)^{\frac{4}{3}}}{20} + c$

6. $\int \frac{1}{4x} dx \quad = \frac{1}{4} \log 4x + c \quad (a = 4, b = 0)$

Alternate method

$\int \frac{1}{4x} dx \quad = \frac{1}{4} \int \frac{1}{x} dx = \frac{1}{4} \log x + c$

7. $\int (5 + x)^{11} dx \quad = \frac{(5+x)^{12}}{12} + c \quad (a = 1, b = 5)$

8. $\int \frac{1}{x+1} dx \quad = \log (x + 1) + c \quad (a = 1, b = 1)$

9. $\int \frac{1}{5-2x} dx \quad = -\frac{1}{2} \log (5 - 2x) + c \quad (a = -2, b = 5)$

10. $\int \frac{1}{(3x+4)^5} dx = \int (3x + 4)^{-5} dx \quad = \frac{(3x+4)^{-4}}{-12} + c = -\frac{1}{12(3x+4)^4} + c$

1.2 Integrals of the form $\int e^x \, dx$, $\int a^x \, dx$ and $\int e^{ax+b} \, dx$

Since $\frac{d}{dx}(e^x) = e^x$ and integration is the inverse process of differentiation, we have

$\int e^x \, dx = e^x + c$

Also $\frac{d}{dx}(a^x) = a^x \log a ==> \int a^x \, dx = \frac{a^x}{\log a} + c$

General form (integrand is a linear function of x)

$\frac{d}{dx}(e^{ax+b}) = e^{ax+b} ==> \int e^{ax+b} \, dx = \frac{1}{a} e^{ax+b} + c$

Remember the following

$$\int e^x \, dx = e^x + c \qquad \int a^x \, dx = \frac{a^x}{\log a} + c$$

$$\int e^{ax+b} \, dx = \frac{1}{a} e^{ax+b} + c$$

Examples 1.2

1. $\int \frac{1}{e^{-x}} dx \quad = \int e^x \, dx = e^x + c$

2. $\int 5^x \, dx \quad = \frac{5^x}{\log 5} + c$

3. $\int 8^x \, dx \quad = \frac{8^x}{\log 8} + c$

4. $\int \frac{1}{e^x} dx \quad = \int e^{-x} \, dx = -e^{-x} + c \qquad (\,a = -1, \ b = 0\,)$

5. $\int e^{2x} \, dx \quad = \frac{1}{2} e^{2x} + c \qquad\qquad (\,a = 2, \quad b = 0\,)$

6. $\int e^{2x+5} \, dx = \frac{1}{2} e^{2x+5} + c \qquad\qquad (\,a = 2, \quad b = 5\,)$

7. $\int e^{3-7x} \, dx = -\frac{1}{7} e^{3-7x} + c \qquad\qquad (\,a = 2, \quad b = 5\,)$

8. $\int \frac{1}{e^{4-x}} dx \quad = \int e^{x-4} \, dx = e^{x-4} + c \qquad (\,a = 1, \ b = -4\,)$

9. $\int (9e^{2x} + 1) \, dx = 9 \int e^{2x} \, dx + \int dx = \frac{9}{2} e^{2x} + x + c$

10. $\int (4^x + x^4 + e^{\frac{x}{5}} + 3) \, dx = \int 4^x \, dx + \int x^4 \, dx + \int e^{\frac{x}{5}} dx + 3 \int dx$

$$= \frac{4^x}{\log 4} + \frac{x^5}{5} + \frac{e^{\frac{x}{5}}}{\frac{1}{5}} + 3x + c = \frac{4^x}{\log 4} + \frac{x^5}{5} + 5 e^{\frac{x}{5}} + 3x + c$$

Exercise 1.2

1. $\int e^{x-3} \, dx \quad = \ldots\ldots\ldots$

2. $\int 4 \, e^{8x-3} dx \quad = \ldots\ldots\ldots$

3. $\int e^{\frac{x-3}{2}} \, dx \quad = \ldots\ldots\ldots$

4. $\int (3e^{3x} - 5) \, dx = \ldots\ldots\ldots$

5. $\int 9^x \, dx \quad = \ldots\ldots\ldots$

Solution 1.2

1. $\int e^{x-3} \, dx \qquad = \ e^{x-3} + c$

2. $\int 4\,e^{8x-3}\,dx \qquad = 4.\dfrac{1}{8}e^{8x-3} + c \;\; = \dfrac{1}{2}e^{8x-3} + c$

3. $\int e^{\frac{x-3}{2}}\,dx \qquad\quad = \int e^{\frac{x}{2}}.e^{-\frac{3}{2}}\,dx \;=\; e^{-\frac{3}{2}}\int e^{\frac{x}{2}}dx \;=\; e^{-\frac{3}{2}}.\dfrac{e^{\frac{x}{2}}}{\frac{1}{2}} + c = 2\,e^{\frac{x-3}{2}} + c$

4. $\int (3e^{3x} - 5)\,dx \quad = 3\int e^{3x}\,dx - 5\int dx = e^{3x} - 5x + c$

5. $\int 9^x\,dx \qquad\qquad = \dfrac{9^x}{\log 9} + c$

1.3 Integrals of the form $\int \sin x\,dx, \int \cos x\,dx$

1. $\dfrac{d}{dx}(\cos x) = -\sin x \;\Longrightarrow\; \int \sin x\,dx = -\cos x + c$

2. $\dfrac{d}{dx}(\sin x) = \cos x \quad \Longrightarrow\; \int \cos x\,dx = \sin x + c$

General form (integrand is a linear function of x)

$\int \sin(ax + b)\,dx = -\dfrac{1}{a}\cos(ax + b) + c$

$\int \cos(ax + b)\,dx = \dfrac{1}{a}\sin(ax + b) + c$

Remember the following

we have found out formulae for $\int \sin x\,dx, \int \cos x\,dx$, only. $\int \tan x\,dx, \int \cot x\,dx, \int \sec x\,dx\ \&\ \int \operatorname{cosec} x\,dx$ **will be discussed later.**
$\int \sin x\,dx = -\cos x + c \qquad\qquad \int \cos x\,dx = \sin x + c$ $\int \sin(ax + b)\,dx = -\dfrac{1}{a}\cos(ax + b) + c$ $\int \cos(ax + b)\,dx = \dfrac{1}{a}\sin(ax + b) + c$

Examples 1.3

1. $\int (2\cos x + 3e^x - 7)dx \;=\; 2\int \cos x\,dx + 3\int e^x\,dx - 7\int dx$

   $\qquad\qquad\qquad\qquad\quad = \; 2\sin x + 3\,e^x - 7x + c$

2. $\int (\sin 3x - 5\cos x - x)dx = \int \sin 3x\,dx - 5\int \cos x\,dx - \int x\,dx$

$$= -\frac{1}{3}\cos 3x - 5\sin x - \frac{x^2}{2} + c$$

3. $\int (9 + \frac{1}{2x+1} - \cos 4x)dx = 9\int dx + \int \frac{1}{2x+1}dx - \int \cos 4x \, dx$

$$= 9x + \frac{1}{2}\log(2x+1) - \frac{1}{4}\sin 4x + c$$

4. $\int (\frac{1}{(5x-2)^4} - \sin 7x)dx = \int \frac{1}{(5x-2)^4}dx - \int \sin 7x \, dx$

$$= \int (5x-2)^{-4} dx - \int \sin 7x \, dx$$

$$= \frac{1}{5} \cdot \frac{(5x-2)^{-3}}{-3} + \frac{1}{7}\cos 7x + c$$

$$= -\frac{1}{15(5x-2)^3} + \frac{1}{7}\cos 7x + c$$

5. $\int (\frac{1}{\cosec 2x} - 5\sqrt{x+1} - \frac{1}{x})dx = \int \sin 2x \, dx - 5\int (x+1)^{\frac{1}{2}} dx - \int \frac{1}{x}dx$

$$= -\frac{\cos 2x}{2} - \frac{10(x+1)^{\frac{3}{2}}}{3} - \log x + c$$

Exercise 1.3

1. $\int (\frac{1}{\sec 2x} - \sin 2x) \, dx = \ldots\ldots\ldots$ 2. $\int \sin(1-5x) \, dx = \ldots\ldots\ldots$

3. $\int (\frac{1}{\cosec 2x} + 2) \, dx = \ldots\ldots\ldots$ 4. $\int \cos(3x+2) \, dx = \ldots\ldots\ldots$

5. $\int \sin x \cos x \, dx = \ldots\ldots\ldots$

Solution 1.3

1. $\int (\frac{1}{\sec 2x} - \sin 2x) \, dx = \int \cos 2x \, dx - \int \sin 2x \, dx = \frac{\sin 2x}{2} + \frac{\cos 2x}{2} + c$

2. $\int \sin(1-5x) \, dx = -\frac{\cos(1-5x)}{-5} + c = \frac{\cos(1-5x)}{5} + c$

3. $\int (\frac{1}{\cosec 2x} + 2) \, dx = \int \sin 2x \, dx + 2\int dx = -\frac{\cos 2x}{2} + 2x + c$

4. $\int \cos(3x+2) \, dx = \frac{\sin(3x+2)}{3} + c$

5. $\int \sin x \cos x \, dx = \frac{1}{2}\int 2\sin x \cos x \, dx = \frac{1}{2}\int \sin 2x \, dx = -\frac{\cos 2x}{2} + c$

1.4 Integrals of the form $\int \sec^2 x\, dx$, $\int \operatorname{cosec}^2 x\, dx$

1. $\dfrac{d}{dx}(\tan x) = \sec^2 x$ $\quad ==> \quad \int \sec^2 x\, dx = \tan x + c$

2. $\dfrac{d}{dx}(\cot x) = -\operatorname{cosec}^2 x \quad ==> \quad \int \operatorname{cosec}^2 x\, dx = -\cot x + c$

General form (integrand is a linear function of x)

$\int \sec^2(ax+b)\, dx = \dfrac{1}{a}\tan(ax+b) + c$

$\int \operatorname{cosec}^2(ax+b)\, dx = -\dfrac{1}{a}\cot(ax+b) + c$

Remember the following

> we have found out formulae for $\int \sec^2 x\, dx$ & $\int \operatorname{cosec}^2 x\, dx$ only.
>
> $\int \sin^2 x\, dx$, $\int \cos^2 x\, dx$, $\int \tan^2 x\, dx$ & $\int \cot^2 x\, dx$
>
> **will be discussed later.**
>
> ---
>
> $\int \sec^2 x\, dx = \tan x + c \qquad \int \operatorname{cosec}^2 x\, dx = -\cot x + c$
>
> $\int \sec^2(ax+b)\, dx = \dfrac{1}{a}\tan(ax+b) + c$
>
> $\int \operatorname{cosec}^2(ax+b)\, dx = -\dfrac{1}{a}\cot(ax+b) + c$

Examples 1.4

1. $\int(4\sec^2 x - 3)\, dx = \int 4\sec^2 x\, dx - 3\int dx = 4\tan x - 3x + c$

2. $\int(\operatorname{cosec}^2 6x + \sqrt{3x+2}\,)\, dx = \int \operatorname{cosec}^2 6x\, dx + \int(3x+2)^{\frac{1}{2}}dx$

$$= \frac{-\cot 6x}{6} + \frac{(3x+2)^{\frac{3}{2}}}{3\cdot\frac{3}{2}} + c = \frac{-\cot 6x}{6} + \frac{2(3x+2)^{\frac{3}{2}}}{9} + c$$

3. $\int \sec^2(5x-3)\, dx = \dfrac{1}{5}\tan(5x-3) + c$

4. $\int(\cos 6x - 7\operatorname{cosec}^2 2x)\, dx = \int \cos 6x\, dx - 7\int \operatorname{cosec}^2 2x\, dx$

$$= \frac{1}{6} \sin 6x + \frac{7}{2} \cot 2x + c$$

5. $\int (\frac{1}{\sin^2(6-5x)} + e^{x+4}) dx = \int \cosec^2(6-5x) \, dx + \int e^{x+4} \, dx$

$$= \frac{1}{5} \cot (6-5x) + e^{x+4} + c$$

Exercise 1.4

1. $\int (\frac{1}{\sin^2 3x} + \sin 2x) \, dx = \ldots\ldots\ldots\ldots$ 2. $\int (\frac{1}{7x-6} + \sec^2 x) \, dx = \ldots\ldots\ldots\ldots$

3. $\int (e^{3x+4} + \frac{1}{\cos^2 5x}) \, dx = \ldots\ldots\ldots\ldots$ 4. $\int \sec^2(4x-7) \, dx = \ldots\ldots\ldots\ldots$

5. $\int \cosec^2(3-2x) \, dx = \ldots\ldots\ldots\ldots$

Solution 1.4

1. $\int (\frac{1}{\sin^2 3x} + \sin 2x) \, dx = \int \cosec^2 3x \, dx + \int \sin 2x \, dx = -\frac{\cot 3x}{3} - \frac{\cos 2x}{2} + c$

2. $\int (\frac{1}{7x-6} + \sec^2 x) \, dx = \int \frac{1}{7x-6} dx + \int \sec^2 x \, dx = \frac{\log(7x-6)}{7} + \tan x + c$

3. $\int (e^{3x+4} + \frac{1}{\cos^2 5x}) \, dx = \int e^{3x+4} \, dx + \int \sec^2 5x \, dx = \frac{e^{3x+4}}{3} + \frac{\tan 5x}{5} + c$

4. $\int \sec^2(4x-7) \, dx = \frac{1}{4} \tan(4x-7) + c$

5. $\int \cosec^2(3-2x) \, dx = \frac{1}{2} \cot(3-2x) + c$

1.5 Integrals of the form $\int \sec x \tan x \, dx, \int \cosec x \cot x \, dx$

1. $\frac{d}{dx}(\sec x) = \sec x \tan x \Rightarrow \int \sec x \tan x \, dx = \sec x + c$

2. $\frac{d}{dx}(\cosec x) = -\cosec x \cot x \Rightarrow \int \cosec x \cot x \, dx = -\cosec x + c$

General form (integrand is a linear function of x)

$\int \sec(ax+b) \tan(ax+b) dx = \frac{1}{a} \sec(ax+b) + c$

$\int \cosec(ax+b) \cot(ax+b) \, dx = -\frac{1}{a} \cosec(ax+b) + c$

Remember the following

$$\int \sec x \tan x \, dx = \sec x + c \qquad \int \mathrm{cosec}\, x \cot x \, dx = -\mathrm{cosec}\ x + c$$

$$\int \sec(ax+b) \tan(ax+b) dx = \frac{1}{a} \sec(ax+b) + c$$

$$\int \mathrm{cosec}(ax+b) \cot(ax+b) \, dx = -\frac{1}{a} \mathrm{cosec}(ax+b) + c$$

Examples 1.5

1. $\int \frac{\sin x}{\cos^2 x} dx \qquad = \int \frac{\sin x}{\cos x} \cdot \frac{1}{\cos x} dx$

$$= \int \tan x \sec x \, dx = \sec x + c$$

2. $\int \frac{\cos x}{\sin^2 x} dx \qquad = \int \frac{\cos x}{\sin x} \cdot \frac{1}{\sin x} dx$

$$= \int \cot x \, \mathrm{cosec}\, x \, dx = -\mathrm{cosec}\ x + c$$

3. $\int \sec(3x+2) \tan(3x+2) \, dx = \frac{1}{3} \sec(3x+2) + c \qquad$ [a = 3, b = 2]

4. $\int \mathrm{cosec}(5-x) \cot(5-x) \, dx = \frac{1}{-1} . - \mathrm{cosec}\ (5-x) + c \quad$ [a = −1, b = 5]

$$= \mathrm{cosec}\ (5-x) + c$$

5. $\int (\cos(\frac{x}{3}+1) - \mathrm{cosec}\ 7x \cot 7x + \frac{1}{e^{2x}}) \, dx$

$$= \int \cos(\frac{x}{3}+1) \, dx - \int \cos ec\ 7x \cot 7x \, dx + \int e^{-2x} \, dx$$

$$= 3 \sin(\frac{x}{3}+1) + \frac{1}{7} \cos ec\ 7x - \frac{1}{2} e^{-2x} + c$$

Exercise 1.5

1. $\int (\sin(2-5x) + \sec\ 2x \tan 2x) \, dx \quad = \, ...$

2. $\int (\mathrm{cosec}\ \frac{2x}{9} \cot \frac{2x}{9} + e^{x+4}) \, dx \qquad = \, ...$

3. $\int (\frac{1}{\cos^2(5x-1)} + \frac{1}{\cos x \cot x}) dx \qquad = \, ...$

4. $\int cosec\,(3x-5)\,\,cot\,(3x-5)\,dx$ = ...

5. $\int sec\,\frac{x}{4}\,\,tan\frac{x}{4}dx$ = ...

Solution 1.5

1. $\int(sin(2-5x)+sec\,2x\,tan\,2x)\,dx$ $= \int sin(2-5x)\,dx + \int sec\,2x\,tan\,2x\,dx$

             $= \frac{1}{5}cos(2-5x)\,+\frac{1}{2}sec\,2\,x + c$

2. $\int(cosec\,\frac{2x}{9}\,cot\,\frac{2x}{9}+e^{x+4})\,dx$ $= \int cosec\,\frac{2x}{9}\,cot\,\frac{2x}{9}\,dx + \int e^{x+4}dx$

             $= -\frac{9}{2}\,cosec\,\frac{2x}{9} + e^{x+4} + c$

3. $\int(\frac{1}{cos^2(5x-1)}+\frac{1}{cos\,x\,cot\,x})dx$ $= \int sec^2(5x-1)\,dx + \int sec\,x\,tan\,x\,dx$

             $= \frac{1}{5}\,tan(5x-1) + sec\,x + c$

4. $\int cosec(3x-5)\,cot(3x-5)\,dx$ $= -\frac{1}{3}\,cosec\,(3x-5) + c$

5. $\int sec\,\frac{x}{4}\,tan\frac{x}{4}\,dx$ $= 4\,sec\,\frac{x}{4} + c$

1.6 Integrals of the form $\int\frac{1}{\sqrt{1-x^2}}dx,\ \int\frac{1}{1+x^2}dx,$

$$\int\frac{1}{1+(ax)^2}dx,\ \int\frac{1}{\sqrt{1-(ax)^2}}dx\ ....$$

1. $\frac{d}{dx}(sin^{-1}x)\,=\,\frac{1}{\sqrt{1-x^2}}\,==>\int\frac{1}{\sqrt{1-x^2}}dx=\,sin^{-1}x+c$

2. $\frac{d}{dx}(tan^{-1}x)\,=\,\frac{1}{1+x^2}\,==>\int\frac{1}{1+x^2}dx\,=\,tan^{-1}x+c$ etc.

General form (integrand is a linear function of x)

$$\int\frac{1}{\sqrt{1-(ax)^2}}dx\,=\,\frac{1}{a}sin^{-1}(ax)+c$$

$$\int\frac{1}{1+(ax)^2}dx\,=\,\frac{1}{a}tan^{-1}(ax)+c\ \text{etc.}$$

Remember the following

$$\int \frac{1}{\sqrt{1-x^2}}\,dx = \sin^{-1}x + c \qquad \int \frac{1}{\sqrt{1-(ax)^2}}\,dx = \frac{1}{a}\sin^{-1}(ax) + c$$

$$\int \frac{1}{\sqrt{1-x^2}}\,dx = -\cos^{-1}x + c \qquad \int \frac{1}{\sqrt{1-(ax)^2}}\,dx = -\frac{1}{a}\cos^{-1}(ax) + c$$

$$\int \frac{1}{1+x^2}\,dx = \tan^{-1}x + c \qquad \int \frac{1}{1+(ax)^2}\,dx = \frac{1}{a}\tan^{-1}(ax) + c$$

$$\int \frac{1}{1+x^2}\,dx = -\cot^{-1}x + c \qquad \int \frac{1}{1+(ax)^2}\,dx = -\frac{1}{a}\cot^{-1}(ax) + c$$

$$\int \frac{1}{x\sqrt{x^2-1}}\,dx = \sec^{-1}x + c \qquad \int \frac{1}{x\sqrt{(ax)^2-1}}\,dx = \frac{1}{a}\sec^{-1}(ax) + c$$

$$\int \frac{1}{x\sqrt{x^2-1}}\,dx = -\csc^{-1}x + c \quad \int \frac{1}{x\sqrt{(ax)^2-1}}\,dx = -\frac{1}{a}\csc^{-1}(ax) + c$$

Examples 1.6

1. $\displaystyle\int \frac{1}{1+2x^2}\,dx = \int \frac{1}{1+(\sqrt{2}x)^2}\,dx$

$$= \frac{1}{\sqrt{2}}\tan^{-1}(\sqrt{2}x) + c \ \ (\text{or}) \ -\frac{1}{\sqrt{2}}\cot^{-1}(\sqrt{2}x) + c$$

2. $\displaystyle\int \frac{1}{\sqrt{1-4x^2}}\,dx = \int \frac{1}{\sqrt{1-(2x)^2}}\,dx$

$$= \frac{1}{2}\sin^{-1}2x + c \ \ (\text{or}) \ -\frac{1}{2}\cos^{-1}2x + c$$

3. $\displaystyle\int \frac{1}{x\sqrt{(5x)^2-1}}\,dx = \frac{1}{5}\sec^{-1}(5x) + c \ \ (\text{or}) \ -\frac{1}{5}\csc^{-1}(5x) + c$

Exercise 1.6

1. $\displaystyle\int \frac{1}{1+\frac{x^2}{9}}\,dx = \ \dots\dots\dots$ 2. $\displaystyle\int \frac{1}{\sqrt{1-3x^2}}\,dx = \ \dots\dots\dots$ 3. $\displaystyle\int \frac{1}{x\sqrt{(7x)^2-1}}\,dx = \ \dots\dots\dots$

Solution 1.6

1. $\displaystyle\int \frac{1}{1+\frac{x^2}{9}}\,dx = \int \frac{1}{1+(\frac{x}{3})^2}\,dx = 3\tan^{-1}\frac{x}{3} + c$

$$(\text{or}) \ \dots\dots\dots\dots \ [\text{try to complete}]$$

2. $\int \frac{1}{\sqrt{1-3x^2}} dx = \int \frac{1}{\sqrt{1-(\sqrt{3}x)^2}} dx$

$= \dots\dots\dots\dots$ (**or**) $-\frac{1}{\sqrt{3}} \cos^{-1} \sqrt{3}x + c$

3. $\int \frac{1}{x\sqrt{7x^2-1}} dx = \int \frac{1}{x\sqrt{(\sqrt{7}x)^2-1}} dx = -\frac{1}{\sqrt{7}} \operatorname{cosec}^{-1} \sqrt{7}x + c$ (**or**) $\dots\dots\dots\dots$

So far we have applied the concept 'integration as the inverse process of differentiation' and evaluated some integrals directly. The difficulty is that this concept cannot be applied to all integrable functions. Various methods are available for this.

Unit 1 Try to recollect the following

		Differentiation	Integration
1.1		$\frac{d}{dx}\left(\frac{x^{n+1}}{n+1}\right) = x^n$	$\int x^n \, dx = \frac{x^{n+1}}{n+1} + c \, , n \neq -1.$
		$\frac{d}{dx}(\log x) = \frac{1}{x}$	$\int x^{-1} \, dx = \int \frac{1}{x} \, dx = \log x + c$
		$\frac{d}{dx}(x) = 1$	$\int dx = x + c$
		$\frac{d}{dx}((ax+b)^{n+1})$	$\int (ax+b)^n \, dx = \frac{1}{a} \cdot \frac{(ax+b)^{n+1}}{n+1} + c$
		$\quad = (n+1)(ax+b)^n . a$	
		$\frac{d}{dx}(\log(ax+b)) = \frac{1}{ax+b} . a$	$\int \frac{1}{ax+b} \, dx = \frac{1}{a} \log(ax+b) + c$
1.2		$\frac{d}{dx}(e^x) = e^x$	$\int e^x \, dx = e^x + c$
		$\frac{d}{dx}(a^x) = a^x \log a$	$\int a^x \, dx = \frac{a^x}{\log a} + c$
		$\frac{d}{dx}(e^{ax+b}) = a \, e^{ax+b}$	$\int e^{ax+b} \, dx = \frac{1}{a} e^{ax+b} + c$
1.3		$\frac{d}{dx}(\sin x) = \cos x$	$\int \cos x \, dx = \sin x + c$
		$\frac{d}{dx}(\cos x) = -\sin x$	$\int \sin x \, dx = -\cos x + c$
		$\frac{d}{dx}(\sin(ax+b)) = a\cos(ax+b)$	$\int \cos(ax+b) \, dx = \frac{1}{a}\sin(ax+b) + c$
		$\frac{d}{dx}(\cos(ax+b)) = -a\sin(ax+b)$	$\int \sin(ax+b) \, dx = -\frac{1}{a}\cos(ax+b) + c$
			Note : $\int \tan x \, dx, \int \cot x \, dx, \int \sec x \, dx$ $\& \int \operatorname{cosec} x \, dx$ **will be discussed later**
1.4		$\frac{d}{dx}(\tan x) = \sec^2 x$	$\int \sec^2 x \, dx = \tan x + c$
		$\frac{d}{dx}(\cot x) = -\operatorname{cosec}^2 x$	$\int \operatorname{cosec}^2 x \, dx = -\cot x + c$
		$\frac{d}{dx}(\tan(ax+b)) = a\sec^2(ax+b)$	$\int \sec^2(ax+b) \, dx = \frac{1}{a}\tan(ax+b) + c$
		$\frac{d}{dx}(\cot(ax+b))$	$\int \operatorname{cosec}^2(ax+b) \, dx$
		$\quad = -a\operatorname{cosec}^2(ax+b)$	$\quad = -\frac{1}{a}\cot(ax+b) + c$
			Note : $\int \sin^2 x \, dx, \int \cos^2 x \, dx, \int \tan^2 x \, dx$ $\& \int \cot^2 x \, dx$ **will be discussed later.**

1.5	$\frac{d}{dx}(\sec x) = \sec x \tan x$	$\int \sec x \tan x \, dx = \sec x + c$
	$\frac{d}{dx}(\cosec x) = -\cosec x \cot x$	$\int \cosec x \cot x \, dx = -\cosec x + c$
	$\frac{d}{dx}(\sec(ax+b))$	$\int \sec(ax+b)\tan(ax+b)dx$
	$= a\sec(ax+b)\tan(ax+b)$	$= \frac{1}{a}\sec(ax+b) + c$
	$\frac{d}{dx}(\cosec(ax+b))$	$\int \cosec(ax+b)\cot(ax+b)\,dx$
	$= -a\cosec(ax+b)\tan(ax+b)$	$= -\frac{1}{a}\cosec(ax+b) + c$
1.6	$\frac{d}{dx}(\sin^{-1}x) = \frac{1}{\sqrt{1-x^2}}$	$\int \frac{1}{\sqrt{1-x^2}}dx = \sin^{-1}x + c$
	$\frac{d}{dx}(\sin^{-1}ax) = \frac{1}{\sqrt{1-(ax)^2}}a$	$\int \frac{1}{\sqrt{1-(ax)^2}}dx = \frac{1}{a}\sin^{-1}(ax) + c$
	$\frac{d}{dx}(\cos^{-1}x) = -\frac{1}{\sqrt{1-x^2}}$	$\int \frac{1}{\sqrt{1-x^2}}dx = -\cos^{-1}x + c$
	$\frac{d}{dx}(\cos^{-1}ax) = -\frac{1}{\sqrt{1-(ax)^2}}a$	$\int \frac{1}{\sqrt{1-(ax)^2}}dx = -\frac{1}{a}\cos^{-1}(ax) + c$
	....	....
	$\frac{d}{dx}(\tan^{-1}x) = \frac{1}{1+x^2}$	$\int \frac{1}{1+x^2}dx = \tan^{-1}x + c$
	$\frac{d}{dx}(\tan^{-1}ax) = \frac{1}{1+(ax)^2}a$	$\int \frac{1}{1+(ax)^2}dx = \frac{1}{a}\tan^{-1}(ax) + c$
	$\frac{d}{dx}(\cot^{-1}x) = -\frac{1}{1+x^2}$	$\int \frac{1}{1+x^2}dx = -\cot^{-1}x + c$
	$\frac{d}{dx}(\cot^{-1}ax) = -\frac{1}{1+(ax)^2}a$	$\int \frac{1}{1+(ax)^2}dx = -\frac{1}{a}\cot^{-1}(ax) + c$
	....	....
	$\frac{d}{dx}(\sec^{-1}x) = \frac{1}{x\sqrt{x^2-1}}$	$\int \frac{1}{x\sqrt{x^2-1}}dx = \sec^{-1}x + c$
	$\frac{d}{dx}(\sec^{-1}ax) = \frac{1}{x\sqrt{(ax)^2-1}}a$	$\int \frac{1}{x\sqrt{(ax)^2-1}}dx = \frac{1}{a}\sec^{-1}(ax) + c$
	$\frac{d}{dx}(\cosec^{-1}x) = -\frac{1}{x\sqrt{x^2-1}}$	$\int \frac{1}{x\sqrt{x^2-1}}dx = -\cosec^{-1}x + c$
	$\frac{d}{dx}(\cosec^{-1}ax) = -\frac{1}{x\sqrt{(ax)^2-1}}a$	$\int \frac{1}{x\sqrt{(ax)^2-1}}dx = -\frac{1}{a}\cosec^{-1}(ax) + c$

After recollecting the formulae (discussed by us in this chapter) given in the above table, try to complete the self evaluation test given below. **You can do it successfully!**

Self Evaluation Test 1

I. Match the following

	Column I			Column II
1	$\int \operatorname{cosec}^2(ax+b)\,dx$	I		$-\dfrac{1}{a}\cot^{-1}(ax)+c$
2	$\dfrac{d}{dx}(\sin^{-1}x)$	II		$\sec x \tan x$
3	$\int \sin x\,dx$	III		$\dfrac{a^x}{\log a}+c$
4	$\dfrac{d}{dx}(\sec x)$	IV		$\dfrac{1}{a}\cdot\dfrac{(ax+b)^{n+1}}{n+1}+c,\quad n\neq -1$
5	$\int \dfrac{1}{x}\,dx$	V		$-\dfrac{1}{a}\cot(ax+b)+c$
6	$\int \dfrac{1}{1+x^2}\,dx$	VI		$-\operatorname{cosec}^{-1}x+c$
7	$\int a^x\,dx$	VII		$\tan^{-1}x+c$
8	$\int \dfrac{1}{x\sqrt{x^2-1}}\,dx$	VIII		$-\cos x+c$
9	$\int \dfrac{1}{1+(ax)^2}\,dx$	IX		$\dfrac{1}{\sqrt{1-x^2}}$
10	$\int (ax+b)^n\,dx$	X		$\log x+c$

Answers

1. V **2.** IX **3.** VIII **4.** II **5.** X **6.** VII **7.** III **8.** VI **9.** I **10.** IV

II Fill in the blanks

1. $\int \dfrac{1}{\sqrt{1-(ax)^2}}\,dx = -\,\ldots\ldots\ldots + c$

2. $\int \ldots\ldots\, dx = -\cos^{-1}x + c$

3. $\dfrac{d}{dx}(\tan^{-1}x) = \cdots \;=> \int \ldots\ldots\, dx = \tan^{-1}x + c$

4. $\int \ldots\ldots\, dx = \dfrac{1}{a}\sec^{-1}(ax) + c$

5. $\int e^{ax+b}\,dx = \ldots\ldots\ldots\, e^{ax+b} + c$

6. $\int \sec(ax+b)\ldots\ldots\ldots\, dx = \dfrac{1}{a}\sec(ax+b) + c$

7. $\dfrac{d}{dx}(\operatorname{cosec} x) = \cdots\ldots\ldots \cot x ==> \int \operatorname{cosec} x \cot x \ldots\ldots = -\operatorname{cosec} x + c$

8. $\int \cos(ax+b)\,dx = \dfrac{1}{a}\ldots\ldots\ldots + c$

9. $\int x^n\,dx = \dfrac{x^{n+1}}{n+1} + c,\ \text{for}\ \ldots\ldots\ldots$

10. $\int \ldots\ldots dx = \frac{1}{a} log(ax + b) + c$

Answers

1. $\frac{1}{a} cos^{-1}(ax)$ 2. $\frac{1}{\sqrt{1-x^2}}$ 3. $\frac{1}{1+x^2}, \frac{1}{1+x^2}$ 4. $\frac{1}{x\sqrt{(ax)^2-1}}$ 5. $\frac{1}{a}$

6. $tan(ax + b)$ 7. $-cosec\ x, dx$ 8. $sin(ax + b)$ 9. $n \neq -1$ 10. $\frac{1}{ax+b}$

III Choose the correct answer

1. $\frac{d}{dx}\left(\frac{x^{n+1}}{n+1}\right) =$ ....

a) x^n b) $\frac{1}{n}x^n$ c) $x^n, \neq -1$ d) x^{n+1}

2. $\int x^{-1} dx =$ .... $+ c$

a) x^{n-1} b) $\frac{1}{x}$ c) e^x d) $log\ x$

3. $\int cos\ x\ dx =$.......... $+ c$

a) $sin\ x$ b) $\frac{1}{x}$ c) $- sin\ x$ d) $log\ x$

4. $\int \ldots\ldots dx = \frac{1}{a} tan(ax + b) + c$

a) $sin(ax + b)$ b) $\frac{1}{(ax+b)}$ c) $sec^2(ax + b)$ d) $log\ (ax + b)$

5. $\frac{d}{dx}(cot^{-1} x) =$ ....

a) $\frac{1}{1+x^2}$ b) $\frac{1}{1-x^2}$ c) $-\frac{1}{1-x^2}$ d) $-\frac{1}{1+x^2}$

Answers

1. a   2. d   3. a   4. c   5. d

IV State True (or) False

1. $\int \frac{1}{x^4} dx = -\frac{1}{3x^3} + c$ 2. $\frac{d}{dx}(\sqrt{3x + 5}) = \frac{2}{3\sqrt{3x+5}}$

3. $\frac{d}{dx}(5^x) = 5^x log\ 5$ 4. $\int sin\ 3x\ dx = \frac{1}{3}cos\ 3x + c$

5. $\int cosec^2\ 3x\ dx = -cot\ 3x + c$ 6. $\int \frac{sin\ x}{cos^2 x} dx = cosec\ x + c$

7. $\int \dfrac{1}{x\sqrt{(ax)^2-1}}\,dx = -\cosec^{-1}(ax) + c$

8. $\int \sec^2 7x\,dx = \dfrac{1}{7}\tan 7x + c$

9. $\int \cosec\dfrac{x}{2}\cot\dfrac{x}{2}\,dx = -2\sec\dfrac{x}{2} + c$

10. $\dfrac{d}{dx}\left(e^{ax+b}\right) = ae^{ax+b}$

Answers

1. T	2. F	3. T	4. F	5. F	6. F

7. F	8. T	9. F	10. T

V. Correct the error if any

1. $\int \dfrac{1}{3x}\,dx = \log 3x + c$

2. $\int dx = x + c$

3. $\int e^{-5x}\,dx = -5\,e^{-5x} + c$

4. $\int \dfrac{1}{\sec 3x}\,dx = -\dfrac{\cos 3x}{3} + c$

5. $\int \sin x \cos x\,dx = -\dfrac{\cos 2x}{4} + c$

6. $\int e^{9x-5}\,dx = 9\,e^{9x-5} + c$

7. $\int \dfrac{1}{\sin x \tan x}\,dx = \cosec x + c$

8. $\dfrac{d}{dx}(\tan^{-1} x) = \dfrac{1}{1-x^2}$

9. $\int \sec^2(5x+6)\,dx = \dfrac{1}{5}\tan(5x+6) + c$

10. $\int \dfrac{1}{\sqrt{1-(ax)^2}}\,dx = \dfrac{1}{a}\cos^{-1}(ax) + c$

Answers

1. $\dfrac{1}{3}\log 3x + c$ (or) $\dfrac{1}{3}\log x + c$ 3. $-\dfrac{1}{5}e^{-5x} + c$ 4. $\dfrac{\sin 3x}{3} + c$ 6. $\dfrac{1}{9}e^{9x-5} + c$

7. $-\cosec x + c$ 8. $\dfrac{1}{1+x^2}$ 10. $\dfrac{1}{a}\sin^{-1}(ax) + c$ (or) $-\dfrac{1}{a}\cos^{-1}(ax) + c$

UNIT 2

Certain functions cannot be integrated directly as in unit 1.

But after some adjustments, these functions can be integrated.

UNIT 2.1 Some functions can be split into **sum and difference of functions** and then integrated
UNIT 2.2 Shortcut methods
UNIT 2.3 Integrals of the form $\int \tan x \, dx, \int \cot x \, dx, \int \sec x \, dx$ & $\int cosecx dx$
UNIT 2.4.1 Integrals of the form $\int \sin^2 x \, dx, \int \cos^2 x \, dx,$ $\int \tan^2 x \, dx, \int \cot^2 x \, dx, \int \sin^4 x \, dx$ & $\int \cos^4 x \, dx$
UNIT 2.4.2 Integrals of the form $\int \sin^3 x \, dx, \int \cos^3 x \, dx$
UNIT 2.4.3 Integrals of the form $\int \sin \alpha x \cos \beta x \, dx$, $\int \sin \alpha x \sin \beta x \, dx$ & $\int \cos \alpha x \cos \beta x \, dx$

2.1 Product and quotient rules are not permitted in integration. The given functions can be split into **sum and difference of functions** and then integrated.

Examples 2.1

1. $\int (1 + x^2)^2 \, dx = \int (1 + x^4 + 2x^2) dx = x + \dfrac{x^5}{5} + \dfrac{2x^3}{3} + c$

2. $\int \dfrac{x^2 + 5x - 3}{x} \, dx = \int x dx + \int 5 \, dx - 3 \int \dfrac{1}{x} dx = \dfrac{x^2}{2} + 5x - 3 \log x + c$

3. $\int (2x - 3)(x + 5) dx = \int (2x^2 + 7x - 15) dx = \dfrac{2x^3}{3} + 7\dfrac{x^2}{2} - 15x + c$

4. $\int \dfrac{e^{2x} + 2e^{-x} - 3}{e^x} dx = \int \dfrac{e^{2x}}{e^x} dx + 2 \int \dfrac{e^{-x}}{e^x} dx - 3 \int \dfrac{1}{e^x} dx$

$\qquad = \int e^x \, dx + 2 \int e^{-2x} \, dx - 3 \int e^{-x} \, dx$

$\qquad = e^x + 2\dfrac{e^{-2x}}{-2} - 3\dfrac{e^{-x}}{-1} + \; = e^x - e^{-2x} + 3e^{-x} + c$

5. $\int (x - 1)\sqrt{x + 4} \, dx = \int \{(x + 4) - 5\} \sqrt{x + 4} \, dx$ \quad [note the adjustment made]

$\qquad = \int (x + 4)\sqrt{x + 4} \, dx - 5 \int \sqrt{x + 4} \, dx$

$\qquad = \int (x + 4)^{\frac{3}{2}} \, dx - 5 \int (x + 4)^{\frac{1}{2}} \, dx$

$$= \frac{(x+4)^{\frac{5}{2}}}{\frac{5}{2}} - \frac{5\,(x+4)^{\frac{3}{2}}}{\frac{3}{2}} + c = \frac{2\,(x+4)^{\frac{5}{2}}}{5} - \frac{10\,(x+4)^{\frac{3}{2}}}{3} + c$$

6. $\int \dfrac{dx}{\sqrt{x+4}-\sqrt{x-2}} = \int \dfrac{\sqrt{x+4}+\sqrt{x-2}}{(\sqrt{x+4}+\sqrt{x-2})(\sqrt{x+4}-\sqrt{x-2})}\,dx$ [note the adjustment made]

$$= \int \frac{\sqrt{x+4}+\sqrt{x-2}}{(x+4-x+2)}\,dx = \frac{1}{6}\int (\sqrt{x+4}+\sqrt{x-2})\,dx$$

$$= \frac{1}{6}\int (x+4)^{\frac{1}{2}}\,dx + \frac{1}{6}\int (x-2)^{\frac{1}{2}}\,dx$$

$$= \frac{1}{6}\frac{(x+4)^{\frac{3}{2}}}{\frac{3}{2}} + \frac{1}{6}\frac{(x-2)^{\frac{3}{2}}}{\frac{3}{2}} + c = \frac{1}{9}(x+4)^{\frac{3}{2}} + \frac{1}{9}(x-2)^{\frac{3}{2}} + c$$

7. $\int e^{x\log 3}.e^x dx = \int e^{\log 3^x}.e^x dx = \int 3^x.e^x dx$

$$= \int (3e)^x\,dx = \frac{(3e)^x}{\log(3e)} + c$$

8. $\int \sqrt{1+\sin 2x}\,dx = \int \sqrt{sin^2x + cos^2x + 2\sin x \cos x}\,dx$

$$= \int \sqrt{(\sin x + \cos x)^2}\,dx = \int (\sin x + \cos x)\,dx$$

$$= -\cos x + \sin x + c$$

9. $\int \dfrac{1}{1-\cos x}\,dx = \int \dfrac{1+\cos x}{(1-\cos x)(1+\cos x)}\,dx = \int \dfrac{1+\cos x}{1-\cos^2 x}\,dx$

$$= \int \frac{1+\cos x}{\sin^2 x}\,dx = \int \frac{1}{\sin^2 x}\,dx + \int \frac{\cos x}{\sin^2 x}\,dx$$

$$= \int cosec^2\,x\,dx + \int \cot x\, cosec x\,dx = -\cot x - cosec x + c$$

10. $\int \dfrac{1+\sin x}{1-\sin x}\,dx = \int \dfrac{(1+\sin x)(1+\sin x)}{(1-\sin x)(1+\sin x)}\,dx = \int \dfrac{(1+\sin x)^2}{\cos^2 x}\,dx$

$$= \int \frac{1}{\cos^2 x}\,dx + \int \frac{\sin^2 x}{\cos^2 x}\,dx + \int \frac{2\sin x}{\cos^2 x}\,dx$$

$$= \int \sec^2 x\,dx + \int \tan^2 x\,dx + \int 2\sec x \tan x\,dx$$

$$= \int \sec^2 x\,dx + \int (\sec^2 x - 1)\,dx + 2\int \sec x \tan x\,dx$$

$$= 2\int \sec^2 x\,dx - \int dx + 2\int \sec x \tan x\,dx$$

$$= 2\tan x - x + 2\sec x + c$$

Exercise 2.1

1. $\int x^2(1+x^3)\,dx = \dots\dots$ 2. $\int (\sqrt{x}-\frac{1}{\sqrt{x}})^2\,dx = \dots\dots$

3. $\int \frac{x^3-x^2+x-1}{x-1}dx$ = ...   4. $\int \frac{1}{1+\sin x}dx$ = ...

5. $\int \frac{x+3}{x\sqrt{x}}dx$ = ...   6. $\int \frac{(1-x^2)^2}{x}dx$ = ...

7. $\int (x+\frac{1}{x})^2 dx$ = ...   8. $\int \frac{1+3e^{2x}}{e^x}dx$ = ...

9. $\int \frac{e^{5x}-e^{-2x}}{e^x}dx$ = ...   10. $\int (e^x+e^{-x})^2 dx$ = ...

Solution 2.1

1. $\int x^2(1+x^3)\ dx = \int(x^2+x^5)dx = \int x^2\ dx+\int x^5\ dx = \frac{x^3}{3}+\frac{x^6}{6}+c$

2. $\int(\sqrt{x}-\frac{1}{\sqrt{x}})^2\ dx = \int(x+\frac{1}{x}-2)\ dx = \frac{x^2}{2}+\log x - 2x+c$

3. $\int \frac{x^3-x^2+x-1}{x-1}dx = \int \frac{x^2(x-1)+(x-1)}{x-1}dx = \int \frac{(x-1)(x^2+1)}{x-1}dx$

$\qquad\qquad = \int(x^2+1)\ dx = \frac{x^3}{3}+x+c$

4. $\int \frac{1}{1+\sin x}dx = \int \frac{1-\sin x}{(1+\sin x)(1-\sin x)}dx = \int \frac{1-\sin x}{1-\sin^2 x}dx$

$\qquad\qquad = \int \frac{1-\sin x}{\cos^2 x}dx = \int \frac{1}{\cos^2 x}dx - \int \frac{\sin x}{\cos^2 x}dx$

$\qquad\qquad = \int \sec^2 x\ dx - \int \tan x\ secx dx$

$\qquad\qquad = \tan x - secx + c$

5. $\int \frac{x+3}{x\sqrt{x}}dx = \int \frac{x}{x\sqrt{x}}dx + 3\int \frac{1}{x\sqrt{x}}dx = \int \frac{1}{x^{\frac{1}{2}}}dx + 3\int \frac{1}{x^{\frac{3}{2}}}dx$

$\qquad\qquad = \int x^{\frac{-1}{2}}dx + 3\int x^{\frac{-3}{2}}dx = 2x^{\frac{1}{2}} - \frac{6}{x^{\frac{1}{2}}} + c$

6. $\int \frac{(1-x^2)^2}{x}dx = \int \frac{1-2x^2+x^4}{x}dx = \int(\frac{1}{x}-2x+x^3)dx$

$\qquad\qquad = \log x - x^2 + \frac{x^4}{4} + c$

7. $\int (x+\frac{1}{x})^2 dx = \int x^2 dx + 2\int dx + \int \frac{1}{x^2}dx = \frac{x^3}{3}+2x-\frac{1}{x}+c$

8. $\int \frac{1+3e^{2x}}{e^x}dx = \int e^{-x}\ dx + 3\int e^x dx = -e^{-x}+3e^x+c$

9. $\int \frac{e^{5x}-e^{-2x}}{e^x}dx = \int e^{4x}\ dx - \int e^{-3x}dx = \frac{e^{4x}}{4}+\frac{e^{-3x}}{3}+c$

10. $\int (e^x + e^{-x})^2\, dx = \int e^{2x}\, dx + 2 \int dx + \int e^{-2x} dx = \dfrac{e^{2x}}{2} + 2x - \dfrac{e^{-2x}}{2} + c$

2.2 Shortcut methods

● **Short cut method 1** : Integrals of the form$\int \dfrac{f'(x)}{f(x)}\, dx$

In the integral $\int \dfrac{f'(x)}{f(x)}\, dx$, the function $f'(x)$ (numerator) is the differential coefficient of the function f(x) (denominator). The formula is $\int \dfrac{f'(x)}{f(x)}\, dx = \log[\text{denominator}] + c$

ie. $\int \dfrac{f'(x)}{f(x)}\, dx = \log[f(x)] + c$ [Try to remember this result]

Examples 2.2.1

1. Consider $\int \dfrac{9}{9x+2}\, dx$

Here $\dfrac{d}{dx}(9x + 2) = 9$ [Note :$\dfrac{d}{dx}$ (denominator) = numerator]

$\Longrightarrow \int \dfrac{9}{9x+2}\, dx = \log$ (denominator) $+ c = \log(9x + 2) + c$

2. $\int \dfrac{1}{x+1}\, dx = \log(x + 1) + c$

3. $\int \dfrac{e^x}{e^x+5}\, dx = \log(e^x + 5) + c$

4. $\int \dfrac{x}{x^2+5}\, dx$ Here to apply the above formula, we require $2x$ in the numerator.

Hence, $\int \dfrac{x}{x^2+5}\, dx = \dfrac{1}{2}\int \dfrac{2x}{x^2+5}\, dx = \dfrac{1}{2}\log(x^2 + 5) + c$ [note the adjustment made]

5. $\int \dfrac{2x-3}{x^2-3x+8}\, dx = \log(x^2 - 3x + 8) + c$

6. $\int \dfrac{\cos x}{1+\sin x}\, dx = \log(1 + \sin x) + c$

7. $\int \dfrac{\sec^2 x}{\tan x}\, dx = \log(\tan x) + c$

8. $\int \dfrac{2\cos x - 3\sin x}{6\cos x + 4\sin x}\, dx = \dfrac{1}{2}\int \dfrac{4\cos x - 6\sin x}{6\cos x + 4\sin x}\, dx = \dfrac{1}{2}\log(6\cos x + 4\sin x) + c$

9. $\int \tan x\, dx = \int \dfrac{\sin x}{\cos x}\, dx = -\int \dfrac{-\sin x}{\cos x}\, dx = -\log(\cos x) + c$

10. $\int \cot x\, dx = \int \dfrac{\cos x}{\sin x}\, dx = \log(\sin x) + c$

Exercise 2.2.1

1. $\int \frac{1}{3x+2} dx$ = ...   2. $\int \frac{1}{e^x+1} dx$ = ...

3. $\int \frac{1}{1+\tan x} dx$ = ...   4. $\int \frac{\sin 2x}{1+\sin^2 x} dx$ = ...

5. $\int \frac{\sin x}{1+\cos x} dx$ = ...   6. $\int \frac{x}{9-4x^2} dx$ = ...

7. $\int \frac{e^{2x}-1}{e^{2x}+1} dx$ = ...   8. $\int \frac{\cos x}{1+\sin x} dx$ = ...

9. $\int \frac{10x^9 + 10^x \log_e 10}{x^{10}+10^x} dx =$ ......   10. $\int \frac{1}{\sin^2 x \cos^2 x} dx$ = ...

Solution 2.2.1

1. $\int \frac{1}{3x+2} dx$ $= \frac{1}{3}\int \frac{3}{3x+2} dx = \frac{1}{3}\log(3x+2) + c$

2. $\int \frac{1}{e^x+1} dx$ $= \int \frac{(e^x+1)-e^x}{e^x+1} dx$

         $= \int \frac{(e^x+1)}{e^x+1} dx - \int \frac{e^x}{e^x+1} dx$

         $= \int dx - \int \frac{e^x}{e^x+1} dx = -\log(e^x+1) + c$

3. $\int \frac{1}{1+\tan x} dx$ $= \int \frac{\cos x}{\cos x+\sin x} dx = \frac{1}{2}\int \frac{2\cos x}{\cos x+\sin x} dx$

         $= \frac{1}{2}\int \frac{(\cos x+\sin x)+(\cos x-\sin x)}{\cos x+\sin x} dx$

         $= \frac{1}{2}\int \frac{\cos x+\sin x}{\cos x+\sin x} dx + \frac{1}{2}\int \frac{\cos x-\sin x}{\cos x+\sin x} dx$

         $= \frac{1}{2}\int dx + \frac{1}{2}\int \frac{-\sin x+\cos x}{\cos x+\sin x} dx$

         $= \frac{1}{2} + \frac{1}{2}\log(\cos x+\sin x) + c$

4. $\int \frac{\sin 2x}{1+\sin^2 x} dx$ $= \int \frac{2\cos x\sin x}{1+\sin^2 x} dx = \log(1+\sin^2 x) + c$

5. $\int \frac{\sin x}{1+\cos x} dx$ $= -\int \frac{-\sin x}{1+\cos x} dx = -\log(1+\cos x) + c$

6. $\int \frac{x}{9-4x^2} dx$ $= -\frac{1}{8}\int \frac{-8x}{9-4x^2} dx = -\frac{1}{8}\log(9-4x^2) + c$

7. $\int \frac{e^{2x}-1}{e^{2x}+1} dx$ Dividing the Nr.(numerator)and the Dr. (denominator) by e^x

         $= \int \frac{e^x-\frac{1}{e^x}}{e^x+\frac{1}{e^x}} dx = \int \frac{e^x-e^{-x}}{e^x+e^{-x}} dx = \log(e^x+e^{-x}) + c$

8. $\int \dfrac{\cos x}{1+\sin x}\,dx \qquad\qquad = \log(1+\sin x) + c$

9. $\int \dfrac{10x^9 + 10^x \log_e 10}{x^{10}+10^x}\,dx = \log(x^{10}+10^x) + c$

10. $\int \dfrac{1}{\sin^2 x\cos^2 x}\,dx = \int \dfrac{\sin^2 x + \cos^2 x}{\sin^2 x\cos^2 x}\,dx$

$$= \int \dfrac{\sin^2 x}{\sin^2 x\cos^2 x}\,dx + \int \dfrac{\cos^2 x}{\sin^2 x\cos^2 x}\,dx$$

$$= \int \sec^2 x\,dx + \int \csc^2 x\,dx = \tan x - \cot x + c$$

● **Shortcut method 2 :** Integrals of the form $\int \dfrac{f'(x)}{\sqrt{f(x)}}\,dx$

In the integral $\int \dfrac{f'(x)}{\sqrt{f(x)}}\,dx$, the function $f'(x)$ (numerator) is the differential coefficient of the function $f(x)$ (function **inside the square root** of the denominator).

The formula is $\int \dfrac{f'(x)}{\sqrt{f(x)}}\,dx = 2\sqrt{f(x)} + c$

Examples 2.2.2

1. $\int \dfrac{\cos x}{\sqrt{\sin x}}\,dx \qquad = 2\sqrt{\sin x} + c \qquad\qquad [\,f(x)=\sin x\,]$

2. $\int \dfrac{e^{2x}}{\sqrt{e^{2x}+3}}\,dx \qquad = \dfrac{1}{2}\int \dfrac{2e^{2x}}{\sqrt{e^{2x}+3}}\,dx = \dfrac{1}{2}\cdot 2.\sqrt{e^{2x}+3} + c = \sqrt{e^{2x}+3} + c$

3. $\int \dfrac{\sin x}{\sqrt{\cos x}}\,dx \qquad = -\int \dfrac{-\sin x}{\sqrt{\cos x}}\,dx = -2\sqrt{\cos x} + c$

4. $\int \dfrac{\sec^2 3x}{\sqrt{\tan 3x}}\,dx \qquad = \dfrac{1}{3}\int \dfrac{3\sec^2 3x}{\sqrt{\tan 3x}}\,dx = \dfrac{2}{3}\sqrt{\tan 3x} + c$

5. $\int \dfrac{\csc 2x \cot 2x}{\sqrt{\csc 2x}}\,dx = -\dfrac{1}{2}\int \dfrac{-2\csc 2x \cot 2x}{\sqrt{\csc 2x}}\,dx = -\sqrt{\csc 2x} + c$

Exercise 2.2.2

1. $\int \dfrac{x+2}{\sqrt{1-x^2}}\,dx \qquad = \ldots\ldots$

2. $\int \dfrac{\sin x}{\sqrt{1-\cos x}}\,dx \qquad = \ldots\ldots$

3. $\int \dfrac{e^x}{\sqrt{e^x+4}}\,dx \qquad = \ldots\ldots$

4. $\int \dfrac{\csc^2 x}{\sqrt{\cot x}}\,dx \qquad = \ldots\ldots$

5. $\int \dfrac{\sec^2 x + \sec x \tan x}{\sqrt{\sec x + \tan x}}\,dx = \ldots\ldots$

Solution 2.2.2

1. $\int \frac{x+2}{\sqrt{1-x^2}} dx \qquad = -\frac{1}{2}\int \frac{-2x}{\sqrt{1-x^2}} dx + 2\int \frac{1}{\sqrt{1-x^2}} dx$

$$= -\frac{1}{2}.2\sqrt{1-x^2} + 2\,sin^{-1}x + c \; = \; -\sqrt{1-x^2} + 2\,sin^{-1}x + c$$

2. $\int \frac{\sin x}{\sqrt{1-\cos x}} dx \qquad = 2\sqrt{1-\cos x} + c$

3. $\int \frac{e^x}{\sqrt{e^x+4}} dx \qquad = 2\sqrt{e^x+4} + c$

4. $\int \frac{cosec^2 x}{\sqrt{\cot x}} dx \qquad = -\int \frac{-cosec^2 x}{\sqrt{\cot x}} dx \; = -2\sqrt{\cot x} + c$

5. $\int \frac{sec^2 x + \sec x \tan x}{\sqrt{\sec x + \tan x}} dx = 2\sqrt{\sec x + \tan x} + c$

● **Shortcut method 3 :** Integrals of the form $\int f'(x)\,[f(x)]^n dx$ where $n \neq -1$

The formula is $\int f'(x)\,[f(x)]^n dx = \dfrac{[f(x)]^{n+1}}{n+1} + c$ **where $n \neq -1$**

Examples 2.2.3

1. $\int \cos x \sin^6 x \, dx \qquad = \frac{\sin^7 x}{7} + c \qquad [f(x) = \sin x]$

2. $\int sec^2 x \tan^5 x \, dx = \int sec^2 x \, [\tan x]^5 \, dx = \frac{\tan^6 x}{6} + c$

3. $\int \sin 2x \cos^3 2x \, dx = -\frac{1}{2}\int - 2\sin 2x \cos^3 2x \, dx = -\frac{\cos^4 2x}{8} + c$

Note : The above three shortcut methods will help us to understand the 'substitution method' (to be discussed in unit3) easily.

Exercise 2.2.3

1. $\int \frac{(\log x)^5}{x} dx \qquad = \ldots\ldots$ 2. $\int \frac{(\tan^{-1} x)^3}{1+x^2} dx \quad = \ldots\ldots$

3. $\int \frac{(1+x)(x+\log x)^5}{x} dx \quad = \ldots\ldots$ 4. $\int \sin x \cos^4 x \, dx \quad = \ldots\ldots$

5. $\int cosec^2 3x \cot^5 3x \, dx \quad = \ldots\ldots$

Solution 2.2.3

1. $\int \frac{(\log x)^5}{x} dx \qquad = \frac{(\log x)^6}{6} + c \qquad [f(x) = \log x]$

2. $\int \dfrac{(\tan^{-1} x)^3}{1+x^2} dx$ $\qquad = \dfrac{(\tan^{-1} x)^4}{4} + c$ $\qquad [f(x) = \tan^{-1} x]$

3. $\int \dfrac{(1+x)(x + \log x)^5}{x} dx$ $\quad = \int (1 + \dfrac{1}{x})\,(x + \log x)^5\, dx\ [\ [f(x) = x + \log x\ ]$

$\qquad\qquad\qquad\qquad = \dfrac{(x + \log x)^6}{6} + c$

4. $\int \sin x \cos^4 x\, dx$ $\qquad = -\int \cos^4 x\,(-\sin x)\, dx\quad [f(x) = \cos x]$

$\qquad\qquad\qquad\qquad = -\dfrac{\cos^5 x}{5} + c$

5. $\int \csc^2 3x \cot^5 3x\, dx$ $\quad = -\dfrac{1}{3}\int \cot^5 3x\,(-3\csc^2 3x)\, dx = -\dfrac{\cot^6 2x}{18} + c$

2.3 Integrals of the form $\int \tan x\, dx, \int \cot x\, dx, \int \sec x\, dx\ \&\ \int \csc x\, dx$

1. $\int \cot x\, dx = \int \dfrac{\cos x}{\sin x} dx = \log(\sin x) + c$ $\qquad [\ \dfrac{d}{dx}(denominator) = numerator]$

2. $\int \tan x\, dx = \int \dfrac{\sin x}{\cos x} dx = -\int \dfrac{-\sin x}{\cos x} dx = -\log(\cos x) + c$

3. $\int \sec x\, dx = \int \dfrac{\sec x\,(\sec x + \tan x)}{\sec x + \tan x} dx$ $\qquad [\ \dfrac{d}{dx}(denominator)=numerator]$

$\qquad\quad = \int \dfrac{\sec^2 x + \sec x \tan x}{\sec x + \tan x} dx$ $\quad = \log(\sec x + \tan x) + c$

4. $\int \csc x\, dx = -\int \dfrac{-\csc x\,(\csc x + \cot x)}{\csc x + \cot x} dx$ $\quad [\ \dfrac{d}{dx}(denominator)=numerator]$

$\qquad\quad = -\int \dfrac{-\csc^2 x - \csc x \cot x}{\csc x + \cot x} dx$ $\quad = -\log(\csc x + \cot x) + c$

Remember the following

$$
\begin{array}{ll}
\int \sin x\, dx & = -\cos x + c \\[4pt]
\int \cos x\, dx & = \sin x + c \\[4pt]
\int \tan x\, dx & = -\log(\cos x) + c \\[4pt]
\int \cot x\, dx & = \log(\sin x) + c \\[4pt]
\int \sec x\, dx & = \log(\sec x + \tan x) + c \\[4pt]
\int \csc x\, dx & = -\log(\csc x + \cot x) + c
\end{array}
$$

Exercise 2.3

1. $\int \tan 3x \, dx =$ ....   2. $\int \cot \frac{x}{2} dx$ = ...

3. $\int \sec 7x \, dx =$ ....   4. $\int cosec \, 4x \, dx$ = ...

Solution 2.3

1. $\int \tan 3x \, dx \qquad = -\frac{1}{3}\int \frac{-3 \sin 3x}{\cos 3x} dx = -\frac{1}{3}\log(\cos 3x) + c$

2. $\int \cot \frac{x}{2} dx \qquad = 2\int \frac{1}{\sin\frac{x}{2}}(\frac{1}{2} \cos \frac{x}{2})dx = 2\log(\sin \frac{x}{2}) + c$

3. $\int \sec 7x \, dx \qquad = \frac{1}{7}\int \frac{7[\sec 7x \, (\sec 7x + \tan 7x)]}{\sec 7x + \tan 7x} dx$

$\qquad\qquad\qquad = \frac{1}{7}\int \frac{7[\sec^2 7x + \sec 7x \tan 7x]}{\sec 7x + \tan 7x} dx$

$\qquad\qquad\qquad = \frac{1}{7}\log(\sec 7x + \tan 7x) + c$

4. $\int cosec \, 4x \, dx \qquad = -\frac{1}{4}\int \frac{-4.cosec \, 4x \, (cosec \, 4x + \cot 4x)}{cosec \, 4x + \cot 4x} dx$

$\qquad\qquad\qquad = -\frac{1}{4}\int \frac{-4[\,cosec^2 \, 4x - cosec \, 4x \cot 4x]}{cosec \, 4x + \cot 4x} dx$

$\qquad\qquad\qquad = -\frac{1}{4}\log(cosec \, 4x + \cot 4x) + c$

2.4 Integration using Trigonometric identities

2.4.1 We use the following trigonometric identities to evaluate $\int \sin^2 x \, dx$,

$\int \cos^2 x \, dx, \int \tan^2 x \, dx, \int \cot^2 x \, dx, \int \sin^4 x \, dx$ and $\int \cos x^4 \, dx$

$1 + tan^2A = sec^2A \qquad\qquad 1 + cot^2A = cosec^2A$

$1 + \cos 2A = 2cos^2A \qquad\qquad 1 + \cos A = 2cos^2\frac{A}{2}$

$1 - \cos 2A = 2sin^2A \qquad\qquad 1 - \cos A = 2sin^2\frac{A}{2}$

Examples 2.4.1

1. $\int \sin^2 x \, dx = \frac{1}{2}\int(1 - \cos 2x) \, dx$

[since $1 - \cos 2x = 2 \sin^2 x, \sin^2 x = \frac{1}{2}(1 - \cos 2x)$]

$\qquad\qquad = \frac{1}{2}\int dx - \frac{1}{2}\int \cos 2x \, dx = \frac{x}{2} - \frac{1}{4} \sin 2x + c$

Complete the following

2. $\int \cos^2 x \, dx = \int \ldots\ldots\ldots\ldots\ldots\ldots\ldots dx \qquad [\, 1 + \cos 2x = 2\cos^2 x \,]$

$\qquad = \ldots\ldots\ldots\ldots\ldots\ldots \qquad = \dfrac{x}{2} + \dfrac{1}{4}\sin 2x + c$

3. $\int \tan^2 x \, dx = \int \ldots\ldots\ldots\ldots\ldots\ldots\ldots dx \qquad [\, 1 + \tan^2 x = \sec^2 x \,]$

$\qquad = \ldots\ldots\ldots\ldots\ldots\ldots \qquad = \tan x - x + c$

4. $\int \cot^2 x \, dx = \int \ldots\ldots\ldots\ldots\ldots\ldots\ldots dx \qquad [\, 1 + \cot^2 x = \csc^2 x \,]$

$\qquad = \ldots\ldots\ldots\ldots\ldots\ldots \qquad = -\cot x - + c$

5. $\int \sin^2 3x \, dx = \dfrac{1}{2}\int (1 - \cos 6x)\, dx$

[since $1 - \cos 2x = 2\sin^2 x$, $\sin^2 x = \dfrac{1}{2}(1 - \cos 2x)$,

$$\sin^2 3x = \dfrac{1}{2}(1 - \cos 6x)]$$

$\qquad = \dfrac{1}{2}\int dx - \dfrac{1}{2}\int \cos 6x \, dx = \dfrac{1}{2}x - \dfrac{1}{12}\sin 6x + c$

6. $\int \cos^2(2x - 1)\, dx$

[since $1 + \cos 2x = 2\cos^2 x$, $\cos^2 x = \dfrac{1}{2}(1 + \cos 2x)$,

$$\cos^2(2x - 1) = \dfrac{1}{2}[1 + \cos(4x - 2)]$$

$\qquad = \dfrac{1}{2}\int dx + \dfrac{1}{2}\int \cos(4x - 2)\, dx$

$\qquad = \dfrac{1}{2}x + \dfrac{1}{8}\sin(4x - 2) + c$

7. $\int \cot^2 5x \, dx \qquad = \int (\csc^2 5x - 1)\, dx \qquad [\, 1 + \cot^2 x = \csc^2 x \,]$

$\qquad\qquad\qquad = -\dfrac{1}{5}\cot 5x - x + c$

8. $\int \tan^2(1 - x)\, dx = \int [\sec^2(1 - x) - 1]\, dx \qquad [\, 1 + \tan^2 x = \sec^2 x \,]$

$\qquad\qquad = \int \sec^2(1 - x)\, dx - \int dx = -\tan(1 - x) - x + c$

9. $\int \sin^4 x \, dx \qquad = \int (\sin^2 x)^2 \, dx = \dfrac{1}{4}\int (1 - \cos 2x)^2 \, dx$

$\qquad\qquad = \dfrac{1}{4}\int (1 - 2\cos 2x + \cos^2 2x)\, dx$

$\qquad\qquad = \dfrac{1}{4}\int [1 - 2\cos 2x + \dfrac{1}{2}(1 + \cos 4x)]\, dx$

$\qquad\qquad = \dfrac{1}{4}\int [\dfrac{3}{2} - 2\cos 2x + \dfrac{1}{2}\cos 4x)]\, dx$

$$= \frac{1}{4} \left[\frac{3}{2}x - 2.\frac{\sin 2x}{2} + \frac{1}{2}.\frac{\sin 4x}{4} \right] + c$$

$$= \frac{3}{8} - \frac{1}{4}\sin 2x + \frac{\sin 4x}{32} + c$$

10. $\int \cos^4 x \, dx$

$$= \frac{1}{4} \int (1 + \cos 2x)^2 \, dx$$

$$= \ldots\ldots\ldots\ldots\ldots\ldots\ldots\ldots\ldots\ldots\ldots\ldots\ldots\ldots$$

$$= \ldots\ldots\ldots\ldots\ldots\ldots\ldots\ldots\ldots\ldots\ldots\ldots\ldots\ldots$$

$$= \frac{1}{4} \int [\frac{3}{2} + 2\cos 2x + \frac{1}{2}\cos 4x) \,] dx$$

$$= \ldots\ldots\ldots\ldots\ldots\ldots\ldots\ldots\ldots\ldots\ldots\ldots\ldots\ldots$$

$$= \frac{3}{8}x + \frac{1}{4}\sin 2x + \frac{\sin 4x}{32} + c$$

2.4.2 Integrals of the form $\int \sin^3 x \, dx, \int \cos^3 x \, dx$ using the identities

$$\sin 3A = 3\sin A - 4\sin^3 A \Longrightarrow \mathbf{\sin^3 A = \frac{1}{4}\,(3\sin A - \sin 3A)}$$

$$\cos 3A = 4\cos^3 A - 3\cos A \Longrightarrow \mathbf{\cos^3 A = \frac{1}{4}\,(3\cos A + \cos 3A)}$$

1. $\int \cos^3 x \, dx = \frac{1}{4} \int (3\cos x + \cos 3x) \, dx$

$$= \frac{3}{4}\sin x + \frac{\sin 3x}{12} + c$$

2. $\int \sin^3 x \, dx = \frac{1}{4} \int (3\sin x - \sin 3x) \, dx$

$$= \frac{-3\cos x}{4} + \frac{\cos 3x}{12} + c$$

3. $\int \sin^3(2x + 1) \, dx = \frac{1}{4} \int [3\sin(2x + 1) - \sin 3(2x + 1)] \, dx$

$$= \frac{3}{4} \int \sin(2x + 1)d - \frac{1}{4} \int \sin(6x + 3) \, dx$$

$$= \frac{-3\cos(2x+1)}{8} + \frac{\cos(6x+3)}{24} + c$$

4. $\int \cos^3(3 - x) \, dx = \frac{1}{4} \int [\cos 3(3 - x) + 3\cos(3 - x)] \, dx$

$$= \frac{1}{4} \int \cos(9 - 3x)dx + \frac{3}{4} \int \cos(3 - x) \, dx$$

$$= \frac{\sin(9-3x)}{-12} + \frac{3}{4}.\frac{\sin(3-x)}{-1} + c$$

$$= -\frac{1}{12}\sin(9 - 3x) - \frac{3}{4}\sin(3 - x) + c$$

5. $\int \sin^3 4x \, dx \qquad = \int \dots\dots\dots\dots\dots\dots. dx$

$\qquad\qquad\qquad = \dfrac{-3\cos 4x}{16} + \dfrac{\cos 12x}{48} + c$

6. $\int \sin^3 2x \, dx \qquad = \int \dots\dots\dots\dots\dots\dots. dx$

$\qquad\qquad\qquad = \dfrac{-3\cos 2x}{8} + \dfrac{\cos 6x}{24} + c$

7. $\int \cos^3 x \, dx \qquad = \frac{1}{4}\int (\cos 3x + 3\cos x)\, dx$

$\qquad\qquad\qquad = \dots\dots\dots\dots\dots\dots\dots + c$

8. $\int \cos^3 5x \, dx \qquad = \frac{1}{4}\int (\cos 15x + 3\cos 5x)\, dx$

$\qquad\qquad\qquad = \dfrac{\sin 15x}{60} + \dfrac{3}{20}\sin 5x + c$

9. $\int \sin^3 (2-x)\, dx \quad = \dots\dots\dots\dots\dots\dots\dots\dots\dots\dots\dots$

$\qquad\qquad\qquad = \dots\dots\dots\dots\dots\dots\dots\dots\dots\dots\dots$

$\qquad\qquad\qquad = \dfrac{3\cos(2-x)}{4} - \dfrac{\cos(6-3x)}{12} + c$

10. $\int \cos^3 (1-3x)\, dx = \dots\dots\dots\dots\dots\dots\dots\dots\dots\dots$

$\qquad\qquad\qquad = \dots\dots\dots\dots\dots\dots\dots\dots\dots\dots\dots$

$\qquad\qquad\qquad = \dots\dots\dots\dots\dots\dots\dots\dots\dots\dots\dots$

$\qquad\qquad\qquad = -\dfrac{1}{36}\sin(3-9x) - \dfrac{1}{4}\sin(1-3x) + c$

2.4.3 Try to recollect the following trigonometric identities

$\sin 2A = 2\sin A \cos A \qquad\qquad \sin A = 2\sin\frac{A}{2}\cos\frac{A}{2}$

$2\sin A \cos B = \sin(A+B) + \sin(A-B)$

$2\cos A \sin B = \sin(A+B) - \sin(A-B)$

$2\cos A \cos B = \cos(A+B) + \cos(A-B)$

$2\sin A \sin B = \cos(A-B) - \cos(A+B)$

$\sin(-A) = -\sin A \qquad \cos(-A) = \cos A$

Examples 2.4.3

1. $\int \sin 4x \cos 4x \, dx = \frac{1}{2}\int 2\sin 4x \cos 4x \, dx$ [Note the adjustment made here]

$$= \frac{1}{2}\int \sin 8x \, dx \qquad [2\sin A \cos A = \sin 2A]$$

$$= \frac{1}{2} \cdot \frac{-\cos 8x}{8} + c = \frac{-\cos 8x}{16} + c$$

2. $\int \sin 5x \cos 3x \, dx = \frac{1}{2}\int 2\sin 5x \cos 3x \, dx$ [Note the adjustment made here]

$$[2\sin A \cos B = \sin(A+B) + sin(A-B)]$$

$$= \frac{1}{2}\int (\sin 8x + \sin 2x) \, dx$$

$$= \frac{1}{2}\left(\frac{-\cos 8x}{8} + \frac{-\cos 2x}{2}\right) + c$$

$$= \frac{-\cos 8x}{16} + \frac{-\cos 2x}{4} + c$$

3. $\int \sin x \cos x \, dx = \frac{1}{2}\int 2\sin x \cos x \, dx$ [Note the adjustment made here]

$$= \frac{1}{2}\int \sin 2x \, dx \qquad [2\sin A \cos A = \sin 2A]$$

$$= \frac{1}{2} \cdot \frac{-\cos 2x}{2} + c = \frac{-\cos 2x}{4} + c$$

4. $\int \sin(2x+5) \cos(2x+5) \, dx = \frac{1}{2}\int 2\sin(2x+5) \cos(2x+5) \, dx$

$$= \frac{1}{2}\int \sin 2(2x+5) \, dx = \frac{1}{2}\int \sin(4x+10) \, dx$$

$$= \frac{1}{2} \cdot \frac{-\cos(4x+10)}{4} + c = \frac{-\cos(4x+10)}{8} + c$$

Exercise 2.4.3

1. $\int \sin 3x \cos 3x \, dx = \ldots\ldots\ldots$ 2. $\int \sin 7x \cos 3x \, dx = \ldots\ldots\ldots$

3. $\int \cos 5x \cos 3x \, dx = \ldots\ldots\ldots$ 4. $\int \sin 2x \cos 5x \, dx = \ldots\ldots\ldots$

5. $\int \sin 9x \sin 2x \, dx = \ldots\ldots\ldots$ 6. $\int \sin x \sin 2x \sin 3x \, dx = \ldots\ldots\ldots$

7. $\int \cos 2x \cos 4x \cos 6x \, dx = \ldots\ldots\ldots$

Solution 2.4.3

1. $\int \sin 3x \cos 3x \, dx = \frac{1}{2}\int 2\sin 3x \cos 3x \, dx$

$$= \frac{1}{2}\int \sin 6x \, dx \qquad [2\sin A \cos A = \sin 2A]$$

$$= \frac{1}{2} \cdot \frac{-\cos 6x}{6} + c = \frac{-\cos 6x}{12} + c$$

2. $\int \sin 7x \cos 3x \, dx = \frac{1}{2} \int 2 \sin 7x \cos 3x \, dx \quad [2 \sin A \cos B = \sin(A+B) + sin(A-B)]$

$$= \frac{1}{2} \int (\sin 10x + \sin 4x) \, dx = \frac{1}{2} \left(\frac{-\cos 10x}{10} + \frac{-\cos 4x}{4} \right) + c$$

$$= -\frac{\cos 10x}{20} - \frac{\cos 4x}{8} + c$$

3. $\int \cos 5x \cos 3x \, dx = \frac{1}{2} \int 2 \cos 5x \cos 3x \, dx \quad [\, 2 \cos A \, \cos B = \cos(A+B) + \cos(A-B)]$

$$= \frac{1}{2} \int (\cos 8x + \cos 2x) \, dx = \frac{1}{2} \left(\frac{\sin 8x}{8} + \frac{\sin 2x}{2} \right) + c$$

$$= \frac{\sin 8x}{16} + \frac{\sin 2x}{4} + c$$

4. $\int \sin 2x \cos 5x \, dx = \frac{1}{2} \int 2 \sin 2x \cos 5x \, dx \quad [2 \sin A \, \cos B = \sin(A+B) + \sin(A-B)]$

$$= \frac{1}{2} \int [\sin 7x + \sin(-3x)] \, dx = \frac{1}{2} \int (\sin 7x - \sin 3x) \, dx$$

$$= \frac{1}{2} \left(\frac{-\cos 7x}{7} - \frac{-\cos 3x}{3} \right) + = \frac{-\cos 7x}{14} + \frac{\cos 3x}{6} + c$$

5. $\int \sin 9x \sin 2x \, dx = \frac{1}{2} \int 2 \sin 9x \sin 2x \, dx \quad [\, 2 \sin A \, \sin B = \cos(A-B) - \cos(A+B)]$

$$= \frac{1}{2} \int (\cos 7x - \cos 11x) \, dx = \frac{1}{2} \left(\frac{\sin 7x}{7} - \frac{\sin 11x}{11} \right) + c$$

$$= \frac{\sin 7x}{14} - \frac{\sin 11x}{22} + c$$

2.5 Evaluation of integrals using 'Partial fractions' will be discussed in Volume II

Unit 2   Try to recollect the following

	Integration
2.1	Some functions can be split into sum and difference of functions and then integrated
2.2	$\int \dfrac{f'(x)}{f(x)} dx = \log[f(x)] + c$ $\qquad$ $\int \dfrac{f'(x)}{\sqrt{f(x)}} dx = 2\sqrt{f(x)} + c$ $\int f'(x)\,[f(x)]^n dx = \dfrac{[f(x)]^{n+1}}{n+1} + c$ $\quad$ where $n \neq -1$
2.3	$\int \tan x\, dx = -\log(\cos x) + c, \quad \int \tan(ax+b)\, dx = -\dfrac{1}{a}\log(\cos(ax+b)) + c$ $\int \cot x\, dx = \log(\sin x) + c, \quad \int \cot(ax+b)\, dx = \dfrac{1}{a}\log(\sin(ax+b)) + c$ $\int \sec x\, dx = \log(\sec x + \tan x) + c$ $\int \sec(ax+b)\, dx = \dfrac{1}{a}\log[\sec(ax+b) + \tan(ax+b)] + c$ $\int \operatorname{cosec} x\, dx = -\log(\operatorname{cosec} x + \cot x) + c$ $\int \operatorname{cosec}(ax+b)\, dx = -\dfrac{1}{a}\log[\operatorname{cosec}(ax+b) + \tan(ax+b)] + c$

2.4.1	$\int \sin^2 x\, dx,$ $\int \cos^2 x\, dx,$ $\int \tan^2 x\, dx, \ \int \cot^2 x\, dx,$ $\int \sin^4 x\, dx, \ \int \cos^4 x\, dx,$	**Trigonometry formulae applied** $1 + \tan^2 A = \sec^2 A \quad ; \quad 1 + \cot^2 A = \operatorname{cosec}^2 A$ $1 - \cos 2A = 2\sin^2 A \ ; \ 1 + \cos 2A = 2\cos^2 A$ $1 + \cos A = 2\cos^2 \dfrac{A}{2} \ ; \ 1 - \cos A = 2\sin^2 \dfrac{A}{2}$
2.4.2	$\int \sin^3 x\, dx,$ $\int \cos^3 x\, dx$	$\sin 3A = 3\sin A - 4\sin^3 A$ $\cos 3A = 4\cos^3 A - 3\cos A$
2.4.3	Integrals of the form $\int \sin \alpha x \cos \beta x\, dx$ $\int \sin \alpha x \sin \beta x\, dx$ $\int \cos \alpha x \cos \beta x\, dx$	$\sin 2A = 2\sin A \cos A \ ; \ \sin A = 2\sin\dfrac{A}{2}\cos\dfrac{A}{2}$ $2\sin A \cos B = \sin(A+B) + \sin(A-B)$ $2\cos A \sin B = \sin(A+B) - \sin(A-B)$ $2\cos A \cos B = \cos(A+B) + \cos(A-B)$ $2\sin A \sin B = \cos(A-B) - \cos(A+B)$ $\sin(-A) = -\sin A \ ; \ \cos(-A) = \cos A$

After recollecting the formulae (discussed by us in this chapter) given in the table, try to complete the self evaluation test given below. **You can do it successfully!**

Self Evaluation Test 2

I. Match the following

	Column I		Column II
1	$\int \frac{f'(x)}{\sqrt{f(x)}} dx$	I	$2 \cos^2 A$
2	$\int \cot x \, dx$	II	$\frac{[f(x)]^{n+1}}{n+1} + c$, $n \neq -1$
3	$\sin 3A$	III	$\cos(A + B) + \cos(A - B)$
4	$1 + \cos 2A$	IV	$2 \sin^2 \frac{A}{2}$
5	$2 \cos A \cos B$	V	$\log(\sec x + \tan x) + c$
6	$\int \tan x \, dx$	VI	$2 \sqrt{f(x)} + c$
7	$\int \sec x \, dx$	VII	$2 \sin \frac{A}{2} \cos \frac{A}{2}$
8	$\sin A$	VIII	$3 \sin A - 4 \sin^3 A$
9	$\int f'(x) \, [f(x)]^n dx$	IX	$\log(\sin x) + c$
10	$1 - \cos A$	X	$-\log(\cos x) + c$

Answers

1. VI **2.** IX **3.** VIII **4.** I **5.** III **6.** X **7.** V **8.** VII **9.** II **10.** IV

II Fill in the blanks

1. $\int \frac{f'(x)}{f(x)} dx = \ldots\ldots\ldots + c$

2. $\int \ldots\ldots \, dx = -\log(\cos x) + c$

3. $\int \sec x \, dx = \log(\sec x + \cdots) + c$

4. $1 + \cdots = cosec^2 A$

5. $1 - \cos 2A = \ldots\ldots\ldots$

6. $\ldots\ldots(3A) = 4 \cos^3 A - 3 \cos A$

7. $2 \sin A \, \sin B = \cos(\ldots\ldots) - \cos(\ldots\ldots)$

8. $\ldots \sin x \cos x = \sin 2x$

9. $\sin \frac{A}{2} \cos \frac{A}{2} = \ldots \sin A$

10. $\ldots\ldots - \cos 2A = 2 \sin^2 A$

Answers

1. $\log[f(x)]$ 2. $\tan x$ 3. $\tan x$ 4. $\cot^2 A$ 5. $2 \sin^2 A$ 6. $\cos$

7. $A - B, A + B$ 8. 2 9. $\frac{1}{2}$ 10. 1

III Choose the correct answer

In the following integrals $(1 - 5)$, which of the following formula is used?

$F_1 : \int \frac{f'(x)}{f(x)} dx = \log[f(x)] + c$ $F_2 : \int \frac{f'(x)}{\sqrt{f(x)}} dx = 2 \sqrt{f(x)} + c$

$F_3 : \int f'(x)[f(x)]^n dx = \dfrac{[f(x)]^{n+1}}{n+1} + c, \ n \neq -1$

1. 1. $\int \cos x \sin^7 x \, dx = \dfrac{\sin^8 x}{8} + c$

a) F_1 b) F_2 c) F_3 d) none of these

2. $\int \dfrac{\cos x}{1+\sin x} dx \quad = \quad \log(1+\sin x) + c$

a) F_1 b) F_2 c) F_3 d) none of these

3. $\int \sec \frac{x}{5} \tan \frac{x}{5} dx \quad = \quad 5\sec \frac{x}{5} + c$

a) F_1 b) F_2 c) F_3 d) none of these

4. $\int \tan x \, dx \quad = \quad -\log(\cos x) + c$

a) F_1 b) F_2 c) F_3 d) none of these

5. $\int \dfrac{e^x}{\sqrt{e^x+2}} dx \quad = \quad 2\sqrt{e^x+2} + c$

a) F_1 b) F_2 c) F_3 d) none of these

Answers   1.c   2. a   3. d   4. a   5. b

IV Problems and their answers are given below. Find out the adjustment to be made here to solve the problems.

Example

$\int \tan x \, dx \ = \ \int \dfrac{\sin x}{\cos x} dx \ = \ -\int \dfrac{-\sin x}{\cos x} dx \ = \ -\log(\cos x) + c$

Adjustment made in the third step. Two negative signs are introduced.

1. $\int \dfrac{x}{x^2-9} dx \ = \ \frac{1}{2} \log(x^2 - 9) + c$

 Adjustment to be made : ....

2. $\int \dfrac{\sin 2x}{1+\sin^2 x} dx \ = \ \log(1+\sin^2 x) + c$

Adjustment to be made : ....

3. $\int \dfrac{\csc^2 x}{\cot x} dx = -\log(\cot x) + c$

Adjustment to be made : ....

4. $\int \dfrac{1+\sin x}{1-\sin x} dx \ = \ 2\tan x - x + 2\sec x + c$

Adjustment to be made : ....

5. $\int \dfrac{dx}{\sqrt{x-4}-\sqrt{x-7}} = \dfrac{2(x-4)^{\frac{3}{2}}}{9} + \dfrac{2(x-7)^{\frac{3}{2}}}{9} + c$

Adjustment to be made : ....

Adjustment made

1. The numerator and the denominator to be multiplied by 2.

2. Adjustment not required.

3. The numerator and the denominator to be multiplied by -1.

4. The numerator and the denominator to be multiplied by $(1 + \sin x)$.

5. The numerator and the denominator to be multiplied by $\sqrt{x - 4} + \sqrt{x - 7}$.

V. Correct the error if any

1. $\int \dfrac{f'(x)}{\sqrt{f(x)}} dx = \sqrt{f(x)} + c$

2. $\int cosec^2 x \, cot^7 x \, dx = -\dfrac{cot^8 x}{8} + c$

3. $\int \dfrac{e^x - e^{-x}}{e^x + e^{-x}} dx = \log(e^x + e^{-x}) + c$

4. $\int f'(x) \, [f(x)]^n dx = \dfrac{[f(x)]^{n+1}}{n+1} + c$

5. $\int \dfrac{(\log x)^9}{x} dx = \dfrac{(\log x)^{10}}{x} + c$

Answers

2, 3 are correct.   1. $2\sqrt{f(x)} + c$ 4. $n \neq -1$ 5. $\dfrac{(\log x)^{10}}{10} + c$

UNIT 3

Integration using substitution

In this method, differentiation plays an important role. The integrand contains more than one function. Among them, a function and its derivative will be there. This function is the required key to evaluate the given integral.

Example 3

1. Consider the integral $\int \frac{\sin{(tan^{-1}x)}}{1+x^2} dx$

clearly $\frac{d}{dx}(tan^{-1}x) = \frac{1}{1+x^2}$.

Hence the substitutionis $t = tan^{-1}x$

Differentiating t with respect to x, we get $\frac{dt}{dx} = \frac{1}{1+x^2} => dt = \frac{1}{1+x^2}dx$

$\therefore \int \frac{\sin{(tan^{-1}x)}}{1+x^2} dx = \int \sin{(tan^{-1}x)}\frac{1}{1+x^2}dx = \int \sin{t}\,dt$

$$= -\cos{t} + c = -\cos{(tan^{-1}x)} + c$$

2. In the integral $\int \frac{(\log{x})^3}{x} dx$, there are two functions **log x** and $\frac{1}{x}$

Which function will be the best choice for the substitution t ?

Clearly $t = \log{x} => \frac{dt}{dx} = \frac{1}{x}$ ie. $dt = \frac{1}{x}dx$. Isn't it?

[Why not we take $t = \frac{1}{x}$? Think it over.]

Hence $\int \frac{(\log{x})^3}{x} dx = \int (\log{x})^3\frac{1}{x}dx = \int t^3\,dt$

$$= \frac{t^4}{4} + c = \frac{(\log{x})^4}{4} + c$$

3. $\int \cot{x}\,dx$

$\int \cot{x}\,dx = \int \frac{\cos{x}}{\sin{x}}dx$

Put $t = \sin{x}$ then $\frac{dt}{dx} = \cos{x}$ ie. $dt = \cos{x}\,dx$

Hence $\int \cot{x}\,dx = \int \frac{\cos{x}}{\sin{x}}dx = \int \frac{1}{t}dt = \log{(t)} + c = \log{(\sin{x})} + c$

[Note : We have already resolved this integral using the formula

$\int \frac{f'(x)}{f(x)} dx = \log[f(x)] + c$ in section 2.2]

4. $\int \frac{3x^2+4x-7}{\sqrt{x^3+2x^2-7x+5}} dx$

Take $t = x^3 + 2x^2 - 7x + 5$ then $\frac{dt}{dx} = (3x^2 + 4x - 7)$ ie. $dt = (3x^2 + 4x - 7) dx$

Hence $\int \frac{3x^2+4x-7}{\sqrt{x^3+2x^2-7x+5}} dx = \int \frac{1}{\sqrt{t}} dt = \int t^{-\frac{1}{2}} dt = 2t^{\frac{1}{2}} + c$

$$= 2\sqrt{(x^3 + 2x^2 - 7x + 5)} + c$$

Alternate method: Using the formula $\int \frac{f'(x)}{\sqrt{f(x)}} dx = 2\sqrt{f(x)} + c$,

we get $\int \frac{3x^2+4x-7}{\sqrt{x^3+2x^2-7x+5}} d = 2\sqrt{(x^3 + 2x^2 - 7x + 5)} + c$

5. $\int x\sqrt{x^2 + 1} dx$

Here we can take either x^2 or $x^2 + 1$ as t. But the best choice will be $x^2 + 1$.

Only then we can come out of the square root easily.

Take $t = x^2 + 1$ then $\frac{dt}{dx} = 2x$ ie. $\frac{1}{2} dt = x dx$

Hence $\int x\sqrt{x^2 + 1} dx = \int \sqrt{x^2 + 1} x dx = \frac{1}{2}\int \sqrt{t} dt$

$$= \frac{1}{2} \cdot \frac{2}{3} t^{\frac{3}{2}} + c \qquad = \frac{1}{3}(x^2 + 1)^{\frac{3}{2}} + c$$

Alternate method: Using the formula $\int f'(x) [f(x)]^n dx = \frac{[f(x)]^{n+1}}{n+1} + c, n \neq -1$

$\int x\sqrt{x^2 + 1} dx = \frac{1}{2}\int 2x(x^2 + 1)^{\frac{1}{2}} dx = \frac{1}{2}\frac{(x^2+1)^{\frac{3}{2}}}{\frac{3}{2}} + c = \frac{1}{3}(x^2 + 1)^{\frac{3}{2}} + c$

6. Consider a) $\int \frac{x^8}{1+x^9} dx$ and b) $\int \frac{x^8}{(1+x^9)^7} dx$

As we have $x^8 dx$ in the numerator of the above two integrals, we can take $t = x^9$.

The first integral can be evaluated easily, but there will be some difficulty in the second integral.

Hence the best choice is $t = 1 + x^9 \Rightarrow \frac{dt}{dx} = 9x^8$ ie. $dt = 9x^8 dx$, $x^8 dx = \frac{1}{9} dt$

$\int \dfrac{x^8}{1+x^9} dx = \int \dfrac{x^8 dx}{1+x^9}$	$\int \dfrac{x^8}{(1+x^9)^7} dx = \int \dfrac{x^8 dx}{(1+x^9)^7}$
$= \dfrac{1}{9} \int \dfrac{dt}{t} = \dfrac{1}{9} \log t + c$	$= \dfrac{1}{9} \int \dfrac{dt}{t^7} = \dfrac{1}{9} \cdot \dfrac{t^{-6}}{-6} + c$
$= \dfrac{1}{9} \log(1+x^9) + c$	$= -\dfrac{1}{54 t^6} + c = -\dfrac{1}{54(1+x^9)^6} + c$
Alternate method - use the formula	Alternate method - use the formula
$\int \dfrac{f'(x)}{f(x)} dx = \log[f(x)] + c$	$\int f'(x)[f(x)]^n dx = \dfrac{[f(x)]^{n+1}}{n+1} + c$

7. $\int \sin^5 x \cos x \, dx$

To get $\cos x \, dx$, let t $= \sin x \Rightarrow \dfrac{dt}{dx} = \cos x$ ie. $dt = \cos x \, dx$

$\int \sin^5 x \cos x \, dx \quad = \int t^5 dt = \dfrac{t^6}{6} + c = \dfrac{1}{6} \sin^6 x + c$

We can use the formula $\int f'(x)[f(x)]^n dx = \dfrac{[f(x)]^{n+1}}{n+1} + c$ also.

8. $\int \dfrac{e^{\cot x}}{\sin^2 x} dx$

$\int \dfrac{e^{\cot x}}{\sin^2 x} dx = -\int e^{\cot x}(-\cosec^2 x)dx$

To get $-\cosec^2 x \, dx$, let $t = \cot x \Rightarrow \dfrac{dt}{dx} = -\cosec^2 x$ ie. $dt = -\cosec^2 x \, dx$

The given integral $= -\int e^t dt = -e^t + c = -e^{\cot x} + c$

Here we can't use the formula $\int f'(x)[f(x)]^n dx = \dfrac{[f(x)]^{n+1}}{n+1} + c$.

9. $\int \dfrac{\sin \sqrt{x}}{\sqrt{x}} dx$

Let $t = \sqrt{x} \Rightarrow \dfrac{dt}{dx} = \dfrac{1}{2\sqrt{x}}$ ie. $dt = \dfrac{1}{2\sqrt{x}} dx$

$\int \dfrac{\sin \sqrt{x}}{\sqrt{x}} dx = 2 \int (\sin \sqrt{x}) \dfrac{1}{2\sqrt{x}} dx = 2 \int \sin t \, dt = -2 \cos t + c = -2 \cos \sqrt{x} + c$

10. $\int \dfrac{\tan^4 \sqrt{x} \sec^2 \sqrt{x}}{\sqrt{x}} dx$

Substitute $t = \tan \sqrt{x} \Rightarrow \dfrac{dt}{dx} = (\sec^2 \sqrt{x}) \dfrac{1}{2\sqrt{x}}$ ie. $dt = \dfrac{\sec^2 \sqrt{x}}{2\sqrt{x}} dx$

$$\int \frac{\tan^4 \sqrt{x} \, \sec^2 \sqrt{x}}{\sqrt{x}} \, dx \;=\; 2 \int \tan^4 \sqrt{x} \cdot \frac{\sec^2 \sqrt{x}}{2\sqrt{x}} \, dx = 2 \int t^4 \, dt$$

$$= \frac{2t^5}{5} + c = \frac{2 \tan^5 \sqrt{x}}{5} + c$$

Alternate method : Here we can use $\int f'(x) \, [f(x)]^n dx = \dfrac{[f(x)]^{n+1}}{n+1} + c$ also.

$$f(x) = \tan \sqrt{x} \;=> f'(x) = \frac{\sec^2 \sqrt{x}}{2\sqrt{x}}$$

$$\int f'(x) \, [f(x)]^n dx \;=\; 2 \int \frac{(\tan \sqrt{x})^4 \, \sec^2 \sqrt{x}}{2 \sqrt{x}} \, dx \;=\; \frac{2 \tan^5 \sqrt{x}}{5} + c$$

10.1 If we have an integral of the form $\int \dfrac{e^{\tan \sqrt{x}} \, \sec^2 \sqrt{x}}{\sqrt{x}} \, dx$, then the above alternate method can't be applied. _Think it over_.

Substitute $t = \tan \sqrt{x}$, then $\dfrac{dt}{dx} = \sec^2 \sqrt{x} \dfrac{1}{2\sqrt{x}}$ ie. $dt = \dfrac{\sec^2 \sqrt{x}}{2\sqrt{x}} \, dx$

Hence $\int \dfrac{e^{\tan \sqrt{x}} \, \sec^2 \sqrt{x}}{\sqrt{x}} \, dx = 2 \int e^{\tan \sqrt{x}} \dfrac{\sec^2 \sqrt{x}}{2\sqrt{x}} \, dx$

$$= 2 \int e^t \, dt = 2 \, e^t + c = 2 \, e^{\tan \sqrt{x}} + c$$

10.2 If we have an integral of the form $\int \dfrac{e^{2 \tan \sqrt{x}} \, \sec^2 \sqrt{x}}{\sqrt{x}} \, dx$, then the above alternate method can also be applied.

To resolve the above integral we substitute $t = e^{\tan \sqrt{x}}$

Then $\dfrac{dt}{dx} = e^{\tan \sqrt{x}} \sec^2 \sqrt{x} \dfrac{1}{2\sqrt{x}}$ ie. $t = e^{\tan \sqrt{x}} \dfrac{\sec^2 \sqrt{x}}{2\sqrt{x}} \, dx$

Hence $\int \dfrac{e^{2 \tan \sqrt{x}} \, \sec^2 \sqrt{x}}{\sqrt{x}} \, dx = 2 \int (e^{\tan \sqrt{x}}) \, e^{\tan \sqrt{x}} \cdot \dfrac{\sec^2 \sqrt{x}}{2\sqrt{x}} \, dx$

$$= 2 \int t \, dt = 2 \frac{t^2}{2} + c = (e^{\tan \sqrt{x}})^2 + c$$

Alternate method - Here we use the formula, $\int [f(x)]^n f'(x) \, dx = \dfrac{[f(x)]^{n+1}}{n+1} + c$

We know that $e^{2 \tan \sqrt{x}} = e^{\tan \sqrt{x}} e^{\tan \sqrt{x}}$

$$f(x) = e^{\tan \sqrt{x}} \;=> f'(x) = e^{\tan \sqrt{x}} \frac{\sec^2 \sqrt{x}}{2\sqrt{x}}$$

$$\int \frac{e^{2\tan\sqrt{x}}\sec^2\sqrt{x}}{\sqrt{x}}\,dx = 2\int (e^{\tan\sqrt{x}})^1\,[e^{\tan\sqrt{x}}\frac{\sec^2\sqrt{x}}{2\sqrt{x}}]\,dx = (e^{\tan\sqrt{x}})^2 + c$$

11. $\int \sin^3 x \cos^2 x\,dx$

$$\int \sin^3 x \cos^2 x\,dx = \int \sin^2 x\,\cos^2 x\,\sin x\,dx = \int (1-\cos^2 x)\,\cos^2 x\,\sin x\,dx$$

Substitute $t = \cos x$, then $\dfrac{dt}{dx} = -\sin x$ ie. $dt = -\sin x\,dx$

Hence $\int \sin^3 x \cos^2 x\,dx \quad = -\int (1-t^2)t^2\,dt \quad = -\int t^2 dt + \int t^4\,dt$

$$= -\frac{t^3}{3} + \frac{t^5}{5} + c \qquad = -\frac{\cos^3 x}{3} + \frac{\cos^5 x}{5} + c$$

12. $\int \dfrac{(x+1)(x+\log x)^2}{x}\,dx \quad = \int \dfrac{(x+1)}{x}(x+\log x)^2\,dx$

$$= \int (x+\log x)^2 \left(1 + \frac{1}{x}\right) dx$$

Substitute $t = x + \log x$, then $dt = \left(1 + \dfrac{1}{x}\right) dx$

Hence $\int \dfrac{(x+1)(x+\log x)^2}{x}\,dx = \int t^2 dt = \dfrac{t^3}{3} + c = \dfrac{(x+\log x)^3}{3} + c$

Alternate method: using the formula, $\int [f(x)]^n f'(x)\,dx = \dfrac{[f(x)]^{n+1}}{n+1} + c,$

we get $\int \dfrac{(x+1)(x+\log x)^2}{x}\,dx = \int (x+\log x)^2 \left(1 + \dfrac{1}{x}\right) dx \quad [f(x) = x + \log x]$

$$= \frac{(x+\log x)^3}{3} + c$$

13. $\int \dfrac{x^3 \sin(\tan^{-1}x^4)}{1+x^8}\,dx$

$$\int \frac{x^3 \sin(\tan^{-1}x^4)}{1+x^8}\,dx = \frac{1}{4}\int \sin(\tan^{-1}x^4)\cdot\frac{4x^3}{1+(x^4)^2}\,dx$$

Substitute $t = \tan^{-1}x^4$, then $dt = \dfrac{1}{1+(x^4)^2}\,4x^3 dx$

The given integral $= \dfrac{1}{4}\int \sin t\,dt = -\dfrac{1}{4}\cos t + c = -\dfrac{1}{4}\cos(\tan^{-1}x^4) + c$

14. $\int e^{2\log x} e^{x^3}\,dx$

We know that $e^{2\log x} = e^{\log x^2} = x^2$. Then $\int e^{2\log x} e^{x^3}\,dx = \int e^{x^3} x^2\,dx$

Take $t = x^3$ then $dt = 3x^2 dx$

Hence $\quad \int e^{2\log x}\, e^{x^3}\, dx \quad = \quad \frac{1}{3}\int e^{x^3}\, 3x^2\, dx = \frac{1}{3}\int e^t\, dt$

$$= \frac{1}{3}e^t + c = \frac{1}{3}e^{x^3} + c$$

15. $\int \tan^5 x\, dx \qquad$ (only odd powers can be resolved in this method)

$\int \tan^5 x\, dx = \int \frac{\sin^5 x}{\cos^5 x}\, dx = \int \frac{\sin^4 x}{\cos^5 x}\sin x\, dx = \int \frac{(1-\cos^2 x)^2}{\cos^5 x}\sin x\, dx$

Take $t = \cos x \quad$ then $\quad dt = -\sin x\, dx$

Hence $\int \tan^5 x\, dx = -\int \frac{(1-t^2)^2}{t^5}\, dt = -\int \frac{1-2t^2+t^4}{t^5}\, dt = -\int \left(\frac{1}{t^5} - \frac{2}{t^3} + \frac{1}{t}\right)dt$

$$= \frac{1}{4t^4} - \frac{1}{t^2} - \log t + c = \frac{1}{4\cos^4 x} - \frac{1}{\cos^2 x} - \log(\cos x) + c$$

Exercise 3

1. $\int \frac{x^5}{1-x^6}\, dx \qquad = \ldots\ldots$

2. $\int \frac{x^2}{\sqrt{1-x^6}}\, dx \qquad = \ldots\ldots$

3. $\int \frac{x^5+4x^2}{\sqrt{1-x^6}}\, dx \qquad = \ldots\ldots$

4. $\int \cos 4x\, dx \qquad = \ldots\ldots$

5. $\int \frac{\sin(\tan x)}{\cos^2 x}\, dx \qquad = \ldots\ldots$

6. $\int x^2\, e^{x^3}\, dx \qquad = \ldots\ldots$

7. $\int \tan x \log \cos x\, dx = \ldots\ldots$

8. $\int \frac{1}{1+\cot x}\, dx \qquad = \ldots\ldots$

9. $\int \frac{\sin x}{(1+\cos x)^5}\, dx \qquad = \ldots\ldots$

10. $\int \sqrt{\cot 3x}\, \csc^2 3x\, dx = \ldots\ldots$

11. $\int \cot^3 x\, dx \qquad = \ldots\ldots$

12. $\int \cos^7 2x\, dx \qquad = \ldots\ldots$

13. $\int \sqrt{\frac{1-\cos x}{1+\cos x}}\, dx \qquad = \ldots\ldots$

14. $\int \tan^4 x\, dx \qquad = \ldots\ldots$

15. $\int \frac{1}{\sin^2 x\cos^2 x}\, dx = \ldots\ldots$

Solution 3

1. $\int \frac{x^5}{1-x^6}\, dx$

As we have x^5 in the numerator, we can take either $x^6 = t$ or $1 - x^6 = t$.

The best choice is $1 - x^6 = t \quad$ Verify.

Then $t = 1 - x^6 \implies \frac{dt}{dx} = -6x^5$ ie. $-\frac{1}{6}dt = x^5 dx$

Hence, $\int \frac{x^5}{1-x^6} dx = -\frac{1}{6}\int \frac{1}{t}dt = -\frac{1}{6}\log t + c = -\frac{1}{6}\log(1 - x^6) + c$

2. $\int \frac{x^2}{\sqrt{1-x^6}} dx$

Here there is only one choice ie. $x^3 = t$.

[The other choice as in the above problem $1 - x^6 = t$ is ruled out. Why?]

Then $t = x^3 \implies \frac{dt}{dx} = 3x^2$ ie. $\frac{1}{3}dt = x^2 dx$

$\int \frac{x^2}{\sqrt{1-x^6}} dx = \frac{1}{3}\int \frac{1}{\sqrt{t}} dt = \frac{1}{3}\int t^{-1/2} dt = \frac{1}{3}\frac{t^{1/2}}{1/2} + c = \frac{2}{3}\sqrt{x^3} + c$

3. $\int \frac{x^5 + 4x^2}{\sqrt{1-x^6}} dx = \int \frac{x^5}{\sqrt{1-x^6}} dx + 4\int \frac{x^2}{\sqrt{1-x^6}} dx$

In the first integral, take $t = 1 - x^6$; In the second integral, take $u = x^3$

Then $\frac{dt}{dx} = -6x^5$ ie. $-\frac{1}{6}dt = x^5 dx$; $\frac{du}{dx} = 3x^2$ ie. $\frac{1}{3}du = x^2 dx$

Hence $\int \frac{x^5 + 4x^2}{\sqrt{1-x^6}} dx = -\frac{1}{6}\int \frac{1}{\sqrt{t}}dt + \frac{4}{3}\int \frac{1}{\sqrt{1-u^2}} du$

$$= -\frac{1}{6}\frac{t^{1/2}}{1/2} + \frac{4}{3}\sin^{-1}u + c = -\frac{1}{3}\sqrt{1-x^6} + \frac{4}{3}\sin^{-1}x^3 + c$$

4. $\int \cos 4x \, dx$

Here take $t = 4x$. Then $\frac{dt}{dx} = 4$; ie. $\frac{1}{4}dt = dx$

Hence $\int \cos 4x \, dx = \frac{1}{4}\int \cos t \, dt = \frac{1}{4}\sin t + c = \frac{1}{4}\sin 4x + c$

Alternate Method : We have already solved this problem using the formula

$$\int \cos(ax + b) \, dx = \frac{1}{a}\sin(ax + b) + c \qquad [\, a = 4, b = 0]$$

5. $\int \frac{\sin(\tan x)}{\cos^2 x} dx = \int \sin(\tan x) . \sec^2 x \, dx$

Here take $t = \tan x$. Then $\frac{dt}{dx} = \sec^2$; ie. $dt = \sec^2 x dx$

Hence $\int \frac{\sin(\tan x)}{\cos^2 x} dx = \int \sin t \, dt = -\cos t + c = -\cos(\tan x) + c$

6. $\int x^2 e^{x^3} dx$

Here there is only one choice   ie. $t = x^3$

Then $t = x^3 \implies \frac{dt}{dx} = 3x^2$ ie. $\frac{1}{3} dt = x^2 dx$

Hence $\int x^2 e^{x^3} dx$ $= \frac{1}{3} \int e^t dt = \frac{1}{3} e^t + c = \frac{1}{3} e^{x^3} + c$

7. $\int \tan x \log \cos x \, dx$ $= \int \frac{\sin x}{\cos x} \log \cos x \, dx$

Here take $t = \cos x$. Then $\frac{dt}{dx} = -\sin x$;   ie. $- dt = \sin x \, dx$

Hence $\int \tan x \log \cos x \, dx = \int \frac{\sin x}{\cos x} \log \cos x \, dx = -\int \frac{1}{t} \log t \, dt$

Again take $u = \log t$. Then $\frac{du}{dt} = \frac{1}{t}$;   ie. $du = \frac{1}{t} dt$

$\int \tan x \log \cos x \, dx$ $= -\int \frac{1}{t} \log t \, d = -\int u \, du$

$$= -\frac{u^2}{2} + c = -\frac{(\log t)^2}{2} + c = -\frac{[\log(\cos x)]^2}{2} + c$$

8. $\int \frac{1}{1+\cot x} dx$ $= \int \frac{\sin x}{\sin x + \cos x} dx = \frac{1}{2} \int \frac{2 \sin x}{\sin x + \cos x} dx$

$$= \frac{1}{2} \int \frac{\sin x + \cos x + \sin x - \cos x}{\sin x + \cos x} dx$$

$$= \frac{1}{2} \int \frac{\sin x + \cos x}{\sin x + \cos x} dx + \frac{1}{2} \int \frac{\sin x - \cos x}{\sin x + \cos x} dx$$

$$= \frac{1}{2} \int dx - \frac{1}{2} \int \frac{-\sin x + \cos x}{\sin x + \cos x} dx \quad [\text{Nr.} = \frac{d}{dx}(Dr)]$$

$$= \frac{1}{2} x - \frac{1}{2} \log(\sin x + \cos x) + c$$

9. $\int \frac{\sin x}{(1+\cos x)^5} dx$ [There are two choices; either $t = \cos x$ or $t = 1 + \cos x$]

The best choice is $t = 1 + \cos x$. Then $\frac{dt}{dx} = -\sin x$; ie. $dt = -\sin x \, dx$

$\int \frac{\sin x}{(1+\cos x)^5} dx = -\int \frac{1}{t^5} dt = -\int t^{-5} dt = \frac{1}{4t^4} + c = \frac{1}{4(1+\cos x)^4} + c$

10. $\int \sqrt{\cot 3x} \, cosec^2 3x \, dx$

Here take $t = \cot 3x$. Then $\frac{dt}{dx} = -3 \, cosec^2 3x$; ie. $-\frac{1}{3} dt = cosec^2 3x \, dx$

Hence $\int \sqrt{\cot 3x}\,cosec^2 3x\,dx \qquad = -\frac{1}{3}\int \sqrt{t}\,dt = -\frac{2}{9}t^{3/2} + c$

$$= -\frac{2}{9}\cot^{3/2}(3x) + c$$

11. $\int \cot^3 x\,dx = \int \frac{\cos^3 x}{\sin^3 x}\,dx = \int \frac{\cos^2 x \cdot \cos x}{\sin^3 x}\,dx = \int \frac{(1-\sin^2 x)\cos x}{\sin^3 x}\,dx$

Here take $t = \sin x$. Then $\frac{dt}{dx} = \cos x$; ie. $dt = \cos x\,dx$

Hence $\int \cot^3 x\,dx \quad = \int \frac{(1-t^2)}{t^3}\,dt = \int \frac{1}{t^3}\,dt - \int \frac{1}{t}\,dt$

$$= -\frac{1}{2\,t^2} - \log\,t + c = -\frac{1}{2\sin^2 x} - \log\,\sin x + c$$

12. $\int \cos^7 2x\,dx \qquad = \int \cos^6 2x \cdot \cos 2x\,dx = \int (1 - \sin^2 2x)^3 \cos 2x\,dx$

Here take $t = \sin 2x$. Then $\frac{dt}{dx} = 2\cos 2x$; ie. $\frac{1}{2}dt = \cos 2x\,dx$

Hence $\int \cos^7 2x\,dx = \frac{1}{2}\int (1-t^2)^3\,dt = \frac{1}{2}\int (1 - 3t^2 + 3t^4 - t^6)\,dt$

$$= \frac{1}{2}\left[t - t^3 + \frac{3t^5}{5} - \frac{t^7}{7}\right] + c$$

$$= \frac{1}{2}\left[\sin 2x - \sin^3 2x + \frac{3\sin^5 2x}{5} - \frac{\sin^7 2x}{7}\right] + c$$

13. $\int \sqrt{\frac{1-\cos x}{1+\cos x}}\,dx \quad = \int \sqrt{\frac{(1-\cos x)(1+\cos x)}{(1+\cos x)(1+\cos x)}}\,dx = \int \sqrt{\frac{1-\cos^2 x}{(1+\cos x)^2}}\,dx$

$$= \int \sqrt{\frac{\sin^2 x}{(1+\cos x)^2}}\,dx = \int \frac{\sin x}{1+\cos x}\,dx$$

Here take $t = 1 + \cos x$. Then $\frac{dt}{dx} = -\sin x$; ie. $-dt = \sin x\,dx$

Hence $\int \sqrt{\frac{1-\cos x}{1+\cos x}}\,dx = -\int \frac{1}{t}\,dt = -\log\,t + c = -\log(1 + \cos x) + c$

14. $\int \tan^4 x\,dx \qquad = \int \tan^2 x \tan^2 x\,dx = \int (\sec^2 x - 1)\tan^2 x\,dx$

$$= \int \sec^2 x \tan^2 x\,dx - \int \tan^2 x\,dx$$

$$= \int \sec^2 x \tan^2 x\,dx - \int (\sec^2 x - 1)\,dx$$

$$= \int \sec^2 x \tan^2 x\,dx - \int \sec^2 x\,dx + \int dx$$

Here take $t = \tan x$. Then $\frac{dt}{dx} = \sec^2 x$; ie. $dt = \sec^2 x\,dx$

$$\int \tan^4 x \, dx \quad = \int t^2 dt - \int \sec^2 x \, dx + \int dx = \frac{t^3}{3} - \tan x + x + c$$

$$= \frac{1}{3}\tan^3 x - \tan x + x + c$$

15. $\int \dfrac{1}{\sin^2 x \cos^2 x} dx = \int \dfrac{\sin^2 x + \cos^2 x}{\sin^2 x \cos^2 x} dx = \int \dfrac{\sin^2 x}{\sin^2 x \cos^2 x} dx + \int \dfrac{\cos^2 x}{\sin^2 x \cos^2 x} dx$

$$= \int \sec^2 x \, dx + \int \text{cosec}^2 x \, dx = \tan x - \cot x + c$$

●━━━━━━━━━━━━━━●━━━━━━━━━━━━━━●

Self Evaluation Test 3

Fill in the blanks with the suitable substitution (SS) and the reason for selecting this to evaluate the given integral

Example

$\int e^x \tan(e^x) \sec(e^x) \, dx$

Put $t = e^x$, then $\dfrac{dt}{dx} = e^x \implies dt = e^x \, dx$

$\int e^x \tan(e^x) \sec(e^x) \, dx = \int \tan t \sec t \, dt = \sec t + c = \sec e^x + c$

Here , SS : $\mathbf{t = e^x}$ Reason : $\dfrac{dt}{dx} = \mathbf{e^x}$ (or) $\mathbf{dt = e^x dx}$

1. $\int \dfrac{\sin(\tan^{-1}x)}{1+x^2} dx$ SS .... Reason ....

2. $\int \dfrac{(\log x)^3}{x} dx$ SS .... Reason ....

3. $\int \cot x \, dx$ SS .... Reason ....

4. $\int \dfrac{x^2}{\sqrt{1-x^6}} dx$ SS .... Reason ....

5. $\int \dfrac{x^5}{(1+x^6)^7} dx$ SS .... Reason ....

6. $\int \sin^7 x \cos x \, dx$ SS .... Reason ....

7. $\int \dfrac{\sin \sqrt{x}}{\sqrt{x}} dx$ SS .... Reason ....

8. $\int \dfrac{e^{\tan \sqrt{x}} \sec^2 \sqrt{x}}{\sqrt{x}} dx$ SS .... Reason ....

9. $\int \frac{(x+1)(x+\log x)^2}{x} dx$ SS .... Reason ....

10. $\int e^{x^3} x^2 dx$ SS .... Reason ....

11. $\int \tan^3 x \, dx$ SS .... Reason ....

12. $\int \frac{\sin(\tan x)}{\cos^2 x} dx$ SS .... Reason ....

13. $\int \sqrt{\cot 3x} \, cosec^2 3x dx$ SS .... Reason ....

14. $\int \frac{x^5}{1+x^{12}} dx$ SS .... Reason ....

15. $\int \frac{x \sin^{-1}(x^2)}{\sqrt{1-x^4}} dx$ SS .... Reason ....

Answers

1. $t = \tan^{-1} x$, $\frac{dt}{dx} = \frac{1}{1+x^2}$ 2. $t = \log x$, $\frac{dt}{dx} = \frac{1}{x}$

3. $t = \sin x$, $\frac{dt}{dx} = \cos x$ 4. $t = x^3$, $\frac{dt}{dx} = 3x^2$

5. $t = 1 + x^6$, $\frac{dt}{dx} = 6x^5$ 6. $t = \sin x$, $\frac{dt}{dx} = \cos x$

7. $t = \sqrt{x}$, $\frac{dt}{dx} = \frac{1}{2\sqrt{x}}$ 8. $t = \tan \sqrt{x}$, $dt = \frac{\sec^2 \sqrt{x}}{2\sqrt{x}} dx$

9. $t = x + \log x$, $dt = \left(1 + \frac{1}{x}\right) dx$ 10. $t = x^3$, $dt = 3x^2 dx$

11. $t = \cos x$, $dt = -\sin x \, dx$ 12. $t = \tan x$, $\frac{dt}{dx} = \sec^2 x$

13. $t = \cot 3x$, $\frac{dt}{dx} = -3 \, cosec^2 3x$ 14. $t = x^6$, $dt = 6 x^5 dx$

15. $t = \sin^{-1}(x^2)$, $\frac{dt}{dx} = \frac{2x}{\sqrt{1-x^4}}$

UNIT 4   Integrals of the form

UNIT 4.1	$\int \frac{1}{\sqrt{x^2-a^2}}\,dx,\ \int \frac{1}{\sqrt{x^2+a^2}}\,dx\ ,\ \int \frac{1}{\sqrt{a^2-x^2}}\,dx$
UNIT 4.2	$\int \frac{1}{x^2-a^2}\,dx,\ \int \frac{1}{x^2+a^2}\,dx\ ,\ \int \frac{1}{a^2-x^2}\,dx$
UNIT 4.3	$\int \frac{1}{ax^2+bx+c}\,dx$ and $\int \frac{1}{\sqrt{ax^2+bx+c}}\,dx$
UNIT 4.4	$\int \frac{px+q}{ax^2+bx+c}\,dx$ and $\int \frac{px+q}{\sqrt{ax^2+bx+c}}\,dx$

Remember the following substitution

For   i. $x^2 - a^2$, substitute $x = a\sec\theta$

ii. $x^2 + a^2$, substitute $x = a\tan\theta$

iii. $a^2 - x^2$, substitute $x = a\sin\theta$

4.1 Integrals of the form $\int \frac{1}{\sqrt{x^2-a^2}}\,dx,\ \int \frac{1}{\sqrt{x^2+a^2}}\,dx\ ,\ \int \frac{1}{\sqrt{a^2-x^2}}\,dx$

1. $\int \frac{1}{\sqrt{x^2-a^2}}\,dx$

Method 1 : $\int \frac{1}{\sqrt{x^2-a^2}}\,dx = \cosh^{-1}\frac{x}{a}+ c$ $[\text{as}\frac{d}{dx}\left(cosh^{-1}x\right) = \frac{1}{\sqrt{x^2-a^2}}]$

Method 2 : Put $x = a\sec\theta$. Then $dx = a\sec\theta\tan\theta\,d\theta$

$\int \frac{1}{\sqrt{x^2-a^2}}\,dx = \int \frac{a\sec\theta\tan\theta}{\sqrt{a^2\sec^2\theta-a^2}}\,d\theta = \int \frac{a\sec\theta\tan\theta}{a\sqrt{(\sec^2\theta-1)}}\,d\theta$

$= \int \frac{\sec\theta\tan\theta}{\tan\theta}\,d\theta = \int \sec\theta\,d\theta$

$\int \frac{1}{\sqrt{x^2-a^2}}\,dx = \log(\sec\theta + \tan\theta) + c$

$x = a\sec\theta => \sec\theta = \frac{x}{a}$

$\tan\theta = \sqrt{\sec^2\theta - 1} = \sqrt{(\frac{x}{a})^2 - 1} = \sqrt{\frac{x^2-a^2}{a^2}} = \frac{\sqrt{x^2-a^2}}{a}$

$\int \frac{1}{\sqrt{x^2-a^2}}\,dx = \log(\sec\theta + \tan\theta) + c = \log(\frac{x}{a} + \frac{\sqrt{x^2-a^2}}{a}) + c$

$$= \log\left(\frac{x + \sqrt{x^2 - a^2}}{a}\right) + c = \log\left(x + \sqrt{x^2 - a^2}\right) - \log a + c$$

$$\int \frac{1}{\sqrt{x^2 - a^2}}\, dx = \log\left(x + \sqrt{x^2 - a^2}\right) + c' \quad \text{where} \quad c' = c - \log a$$

Thus, $\int \frac{1}{\sqrt{x^2 - a^2}}\, dx = \cosh^{-1}\frac{x}{a} + c$ (or) $\log\left(x + \sqrt{x^2 - a^2}\right) + c'$

2. $\int \frac{1}{\sqrt{x^2 + a^2}}\, dx$

Method 1 : $\int \frac{1}{\sqrt{x^2 + a^2}}\, dx = \sinh^{-1}\frac{x}{a} + c$ $\quad \left[\text{as } \frac{d}{dx}\left(\sinh^{-1}x\right) = \frac{1}{\sqrt{x^2 + a^2}}\right]$

Method 2 : Put $x = a\tan\theta$. Then $dx = a\sec^2\theta\, d\theta$

$$\int \frac{1}{\sqrt{x^2 + a^2}}\, dx = \int \frac{a\sec^2\theta}{\sqrt{a^2\tan^2\theta + a^2}}\, dx = \int \frac{a\sec^2\theta}{a\sqrt{(\tan^2\theta + 1)}}\, d\theta = \int \sec\theta\, d\theta$$

ie. $\int \frac{1}{\sqrt{x^2 + a^2}}\, dx = \log(\sec\theta + \tan\theta) + c$

$x = a\tan\theta \Rightarrow \tan\theta = \frac{x}{a}$,

then $\sec\theta = \sqrt{1 + \tan^2\theta} = \sqrt{1 + \left(\frac{x}{a}\right)^2} = \sqrt{\frac{x^2 + a^2}{a^2}} = \frac{\sqrt{x^2 + a^2}}{a}$

$$\int \frac{1}{\sqrt{x^2 + a^2}}\, dx = \log(\sec\theta + \tan\theta) + c$$

$$= \log\left(\frac{x}{a} + \frac{\sqrt{x^2 + a^2}}{a}\right) + c = \log\left(\frac{x + \sqrt{x^2 + a^2}}{a}\right) + c$$

$$= \log\left(x + \sqrt{x^2 + a^2}\right) - \log a + c$$

$$\int \frac{1}{\sqrt{x^2 + a^2}}\, dx = \log\left(x + \sqrt{x^2 + a^2}\right) + c' \quad \text{where} \quad c' = c - \log a$$

Thus, $\int \frac{1}{\sqrt{x^2 + a^2}}\, dx = \sinh^{-1}\frac{x}{a} + c$ (or) $\log\left(x + \sqrt{x^2 + a^2}\right) + c'$

3. $\int \frac{1}{\sqrt{a^2 - x^2}}\, dx$ $\quad$ Put $x = a\sin\theta$. Then $dx = a\cos\theta\, d\theta$

$$\int \frac{1}{\sqrt{a^2 - x^2}}\, dx = \int \frac{a\cos\theta}{\sqrt{a^2 - a^2\sin^2\theta}}\, d\theta = \int \frac{a\cos\theta}{a\sqrt{(1 - \sin^2\theta)}}\, d\theta = \int d\theta$$

$$= \theta + c = \sin^{-1}\frac{x}{a} + c \quad [x = a\sin\theta \Rightarrow \theta = \sin^{-1}\frac{x}{a}]$$

$$\int \frac{1}{\sqrt{a^2 - x^2}}\, dx = \sin^{-1}\frac{x}{a} + c$$

Remember the following

$$\int \frac{1}{\sqrt{x^2-a^2}}\, dx = \cosh^{-1}\frac{x}{a} + c \ \text{(or)}\ \log(x + \sqrt{x^2 - a^2}) + c'$$

$$\int \frac{1}{\sqrt{x^2+a^2}}\, dx = \sinh^{-1}\frac{x}{a} + c \ \text{(or)}\ \log(x + \sqrt{x^2 + a^2}) + c'$$

$$\int \frac{1}{\sqrt{a^2-x^2}}\, dx = \sin^{-1}\frac{x}{a} + c$$

Examples 4.1

1. $\int \dfrac{1}{\sqrt{16-x^2}}\, dx = \int \dfrac{1}{\sqrt{4^2-x^2}}\, dx = \sin^{-1}\dfrac{x}{4} + c$

2. $\int \dfrac{1}{\sqrt{9-4x^2}}\, dx = \dfrac{1}{2}\int \dfrac{1}{\sqrt{(3/_2)^2-x^2}}\, dx = \dfrac{1}{2}\sin^{-1}\dfrac{2x}{3} + c$

3. $\int \dfrac{1}{\sqrt{16+x^2}}\, dx = \int \dfrac{1}{\sqrt{4^2+x^2}}\, dx = \sinh^{-1}\dfrac{x}{4} + c\ \text{(or)}\ \log(x + \sqrt{x^2 + 16}) + c'$

4. $\int \dfrac{1}{\sqrt{9+4x^2}}\, dx \qquad = \dfrac{1}{2}\int \dfrac{1}{\sqrt{(3/_2)^2+x^2}}\, dx$

$$= \dfrac{1}{2}\sinh^{-1}\dfrac{2x}{3} + c \quad \text{(or)}\ \dfrac{1}{2}\log(2x + \sqrt{4x^2 + 9}) + c'$$

5. $\int \dfrac{1}{\sqrt{x^2-16}}\, dx \qquad = \int \dfrac{1}{\sqrt{x^2- 4^2}}\, dx$

$$= \cosh^{-1}\dfrac{x}{4} + c \qquad \text{(or)}\ \log(x + \sqrt{x^2 - 16}) + c'$$

6. $\int \dfrac{1}{\sqrt{4x^2-9}}\, dx \qquad = \dfrac{1}{2}\int \dfrac{1}{\sqrt{x^2- (3/_2)^2}}\, dx$

$$= \dfrac{1}{2}\cosh^{-1}\dfrac{2x}{3} + c \quad \text{(or)}\ \dfrac{1}{2}\log(2x + \sqrt{4x^2 - 9}) + c'$$

7. $\int \dfrac{1}{\sqrt{a^2-b^2x^2}}\, dx = \dfrac{1}{b}\int \dfrac{1}{\sqrt{(a/_b)^2-x^2}}\, dx = \dfrac{1}{b}\sin^{-1}\dfrac{bx}{a} + c$

8. $\int \dfrac{sec^2x}{\sqrt{tan^2x-16}}\, dx$

Put $t = \tan x$. Then $dt = \sec^2 x\, dx$

$$\int \frac{sec^2x}{\sqrt{tan^2x-16}}\, dx = \int \frac{dt}{\sqrt{t^2- 4^2}} = \cosh^{-1}\frac{t}{4} + c$$

$$(or) \log (t + \sqrt{t^2 - 16}) + c' \text{ where } t = \tan x$$

Exercises 4.1

1. $\int \frac{1}{\sqrt{9-x^2}} dx$ $\quad = \quad \dots\dots$ 2. $\int \frac{1}{\sqrt{4-9x^2}} dx$ $\quad = \quad \dots\dots$

3. $\int \frac{1}{\sqrt{25+x^2}} dx$ $\quad = \quad \dots\dots$ 4. $\int \frac{1}{\sqrt{16+9x^2}} dx$ $\quad = \quad \dots\dots$

5. $\int \frac{1}{\sqrt{x^2-64}} dx$ $\quad = \quad \dots\dots$ 6. $\int \frac{1}{\sqrt{9x^2-4}} dx$ $\quad = \quad \dots\dots$

7. $\int \frac{1}{\sqrt{l^2-m^2x^2}} dx$ $\quad = \quad \dots\dots$ 8. $\int \frac{1}{\sqrt{l-mx^2}} dx$ $\quad = \quad \dots\dots$

Solutions 4.1

1. $\int \frac{1}{\sqrt{9-x^2}} dx$ $\quad = \quad \int \frac{1}{\sqrt{3^2-x^2}} dx \quad = \sin^{-1}\frac{x}{3} + c$

2. $\int \frac{1}{\sqrt{4-9x^2}} dx$ $\quad = \frac{1}{3}\int \frac{1}{\sqrt{(^2/_3)^2 - x^2}} dx = \frac{1}{3}\sin^{-1}\frac{3x}{2} + c$

3. $\int \frac{1}{\sqrt{25+x^2}} dx$ $\quad = \quad \int \frac{1}{\sqrt{5^2+x^2}} dx \quad = \sinh^{-1}\frac{x}{5} + c$

$$(or) \log (x + \sqrt{x^2 + 25}) + c'$$

4. $\int \frac{1}{\sqrt{16+9x^2}} dx$ $\quad = \frac{1}{3}\int \frac{1}{\sqrt{(^4/_3)^2 + x^2}} dx \quad = \frac{1}{3}\sinh^{-1}\frac{3x}{4} + c$

$$(or) \frac{1}{3} \log (3x + \sqrt{9x^2 + 16}) + c'$$

5. $\int \frac{1}{\sqrt{x^2-64}} dx$ $\quad = \quad \int \frac{1}{\sqrt{x^2 - 8^2}} dx \quad = \cosh^{-1}\frac{x}{8} + c$

$$(or) \log (x + \sqrt{x^2 - 64}) + c'$$

6. $\int \frac{1}{\sqrt{9x^2-4}} dx$ $\quad = \frac{1}{3}\int \frac{1}{\sqrt{x^2 - (^2/_3)^2}} dx \quad = \frac{1}{3}\cosh^{-1}\frac{3x}{2} + c$

$$(or) \frac{1}{3} \log (3x + \sqrt{9x^2 - 4}) + c'$$

7. $\int \frac{1}{\sqrt{l^2-m^2x^2}} dx$ $\quad = \frac{1}{m}\int \frac{1}{\sqrt{(^l/_m)^2 - x^2}} dx \quad = \frac{1}{m}\sin^{-1}\frac{mx}{l} + c$

8. $\int \frac{1}{\sqrt{l-mx^2}} dx \quad = \frac{1}{\sqrt{m}} \int \frac{1}{\sqrt{(\sqrt{l/m})^2 - x^2}} dx = \frac{1}{\sqrt{m}} \sin^{-1} \frac{\sqrt{m}}{\sqrt{l}} x + c$

4. 2 Integrals of the form $\int \frac{1}{x^2-a^2} dx, \int \frac{1}{x^2+a^2} dx, \int \frac{1}{a^2-x^2} dx$

We use the following expression in the evaluation of the first two integrals

$$\frac{1}{x^2-a^2} = \frac{1}{(x-a)(x+a)} = \frac{1}{2a} [\frac{1}{x-a} - \frac{1}{x+a}]$$

$$\frac{1}{a^2-x^2} = \frac{1}{(a-x)(a+x)} = \frac{1}{2a} [\frac{1}{a-x} + \frac{1}{a+x}]$$

1. $\int \frac{1}{x^2-a^2} dx$

$\int \frac{1}{x^2-a^2} dx \quad = \int \frac{1}{(x-a)(x+a)} dx = \frac{1}{2a} \int [\frac{1}{x-a} - \frac{1}{x+a}] dx$

$$= \frac{1}{2a} [\int \frac{1}{x-a} dx - \int \frac{1}{x+a} dx] = \frac{1}{2a}[\log(x-a) - \log(x+a)] + c$$

$\int \frac{1}{x^2-a^2} dx \quad = \frac{1}{2a} \log \left(\frac{x-a}{x+a}\right) + c$

2. $\int \frac{1}{a^2-x^2} dx$

$\int \frac{1}{a^2-x^2} dx \quad = \int \frac{1}{(a-x)(a+x)} dx = \frac{1}{2a} \int [\frac{1}{a-x} + \frac{1}{a+x}] dx$

$$= \frac{1}{2a} [\int \frac{1}{a-x} dx + \int \frac{1}{a+x} dx] = \frac{1}{2a}[- \log(a-x) + \log(a+x)] + c$$

$\int \frac{1}{a^2-x^2} dx \quad = \frac{1}{2a} \log(\frac{a+x}{a-x}) + c$

3. $\int \frac{1}{x^2+a^2} dx$

Put $x = a \tan\theta$. Then $dx = a \sec^2\theta\, d\theta$

$\int \frac{1}{x^2+a^2} dx = \int \frac{a \sec^2\theta}{a^2 \tan^2\theta + a^2} d\theta \quad = \frac{1}{a} \int \frac{\sec^2\theta}{\sec^2\theta} d\theta$

$$= \frac{1}{a} \int d\theta = \frac{1}{a}\theta + c \quad [x = a \tan\theta => \theta = \tan^{-1}\frac{x}{a}]$$

$\int \frac{1}{x^2+a^2} dx = \frac{1}{a} \tan^{-1}\frac{x}{a} + c$

Remember the following

$$\int \frac{1}{x^2-a^2}\, dx = \frac{1}{2a} \log\left(\frac{x-a}{x+a}\right) + c$$

$$\int \frac{1}{a^2-x^2}\, dx = \frac{1}{2a} \log\left(\frac{a+x}{a-x}\right) + c$$

$$\int \frac{1}{x^2+a^2}\, dx = \frac{1}{a} \tan^{-1}\frac{x}{a} + c$$

Examples 4.2

1. $\int \frac{1}{x^2-16}\, dx \quad = \int \frac{1}{x^2-4^2}\, dx = \frac{1}{8}\log\left(\frac{x-4}{x+4}\right) + c \quad$ [Here $a=4$]

2. $\int \frac{1}{4x^2-25}\, dx \quad = \frac{1}{4}\int \frac{1}{x^2-\frac{25}{4}}\, dx = \frac{1}{4}\int \frac{1}{x^2-\left(\frac{5}{2}\right)^2}\, dx \quad$ [Here $a=\frac{5}{2}$]

$$= \frac{1}{20}\log\left(\frac{x-\frac{5}{2}}{x+\frac{5}{2}}\right) + c = \frac{1}{20}\log\left(\frac{2x-5}{2x+5}\right) + c$$

3. $\int \frac{1}{9-x^2}\, dx \quad = \int \frac{1}{3^2-x^2}\, dx \quad = \frac{1}{6}\log\left(\frac{3+x}{3-x}\right) + c \quad$ [Here $a=3$]

4. $\int \frac{1}{9-16x^2}\, dx \quad = \frac{1}{16}\int \frac{1}{\frac{9}{16}-x^2}\, dx = \frac{1}{16}\int \frac{1}{\left(\frac{3}{4}\right)^2-x^2}\, dx$

$$= \frac{1}{24}\log\left(\frac{\frac{3}{4}+x}{\frac{3}{4}-x}\right) + c = \frac{1}{24}\log\left(\frac{3+4x}{3-4x}\right) + c$$

5. $\int \frac{1}{x^2+36}\, dx \quad = \int \frac{1}{x^2+6^2}\, dx \quad = \frac{1}{6}\tan^{-1}\frac{x}{6} + c \quad$ [Here $a=6$]

6. $\int \frac{1}{16x^2+9}\, dx \quad = \frac{1}{16}\int \frac{1}{x^2+\frac{9}{16}}\, dx \quad = \frac{1}{16}\int \frac{1}{x^2+\left(\frac{3}{4}\right)^2}\, dx \quad$ [Here $a=\frac{3}{4}$]

$$= \frac{1}{12}\tan^{-1}\frac{4x}{3} + c$$

Exercise 4.2

1. $\int \frac{1}{x^2-64}\, dx \quad = \quad \ldots\ldots\ldots$

2. $\int \frac{1}{4x^2-5}\, dx \quad = \quad \ldots\ldots\ldots$

3. $\int \frac{1}{16-x^2}\, dx \quad = \quad \ldots\ldots\ldots$

4. $\int \frac{1}{4-25x^2}\, dx \quad = \quad \ldots\ldots\ldots$

5. $\int \frac{1}{x^2+3}\, dx \quad = \quad \ldots\ldots\ldots$

6. $\int \frac{1}{25x^2+4}\, dx \quad = \quad \ldots\ldots\ldots$

Solution 4.2

1. $\int \frac{1}{x^2-64} dx \qquad = \int \frac{1}{x^2-8^2} dx = \frac{1}{16} \log\left(\frac{x-8}{x+8}\right) + c$ [Here $a = 8$]

2. $\int \frac{1}{4x^2-5} dx \qquad = \frac{1}{4} \int \frac{1}{x^2-\frac{5}{4}} dx = \frac{1}{4} \int \frac{1}{x^2-(\frac{\sqrt{5}}{2})^2} dx$ [Here $a = \frac{\sqrt{5}}{2}$]

$$= \frac{1}{4\sqrt{5}} \log\left(\frac{x-\frac{\sqrt{5}}{2}}{x+\frac{\sqrt{5}}{2}}\right) + c = \frac{1}{4\sqrt{5}} \log\left(\frac{2x-\sqrt{5}}{2x+\sqrt{5}}\right) + c$$

3. $\int \frac{1}{16-x^2} dx \qquad = \int \frac{1}{4^2-x^2} dx = \frac{1}{8} \log\left(\frac{4+x}{4-x}\right) + c$ [Here $a = 4$]

4. $\int \frac{1}{4-25x^2} dx \qquad = \frac{1}{25} \int \frac{1}{\frac{4}{25}-x^2} dx \qquad = \frac{1}{25} \int \frac{1}{(\frac{2}{5})^2-x^2} dx$ [Here $a = \frac{2}{5}$]

$$= \frac{1}{20} \log\left(\frac{\frac{2}{5}+x}{\frac{2}{5}-x}\right) + c = \frac{1}{20} \log\left(\frac{2+5x}{2-5x}\right) + c$$

5. $\int \frac{1}{x^2+3} dx \qquad = \int \frac{1}{x^2+(\sqrt{3})^2} dx \qquad = \frac{1}{\sqrt{3}} \tan^{-1} \frac{x}{\sqrt{3}} + c$ [Here $a = \sqrt{3}$]

6. $\int \frac{1}{25x^2+4} dx \qquad = \frac{1}{25} \int \frac{1}{x^2+\frac{4}{25}} dx \qquad = \frac{1}{25} \int \frac{1}{x^2+(\frac{2}{5})^2} dx$ [Here $a = \frac{2}{5}$]

$$= \frac{1}{10} \tan^{-1} \frac{5x}{2} + c$$

4.3 Integrals of the form $\int \frac{1}{ax^2+bx+c} dx$ and $\int \frac{1}{\sqrt{ax^2+bx+c}} dx$

4.3.1 $\int \frac{1}{ax^2+bx+c} dx$

Formulae required to evaluate this type of problems

$$\int \frac{1}{x^2-a^2} dx = \frac{1}{2a} \log\left(\frac{x-a}{x+a}\right) + c ; \qquad \int \frac{1}{a^2-x^2} dx = \frac{1}{2a} \log\left(\frac{a+x}{a-x}\right) + c$$

$$\int \frac{1}{x^2+a^2} dx = \frac{1}{a} \tan^{-1} \frac{x}{a} + c$$

Examples 4.3.1

1. $\int \frac{1}{x^2+6x+5} dx$

Method :We know that $a^2 + 2ab + b^2 = (a + b)^2$

Consider $x^2 + 6x + 5$

By comparing $x^2 + 6x$ with $a^2 + 2ab$, we find $a = x$; $2ab = 6x => b = 3$

Now $x^2 + 6x + 5 = \underline{x^2 + 2.3.x + 3^2 - 3^2 + 5}$ [add and subtract b^2 ie 3^2]

$x^2 + 6x + 5 = (x + 3)^2 - 2^2$

$\int \frac{1}{x^2+6x+5} dx = \int \frac{1}{(x+3)^2-2^2} dx = \frac{1}{4} \log \left(\frac{x+3-2}{x+3+2} \right) + c = \frac{1}{4} \log \left(\frac{x+1}{x+5} \right) + c$

2. $\int \frac{1}{x^2+2x+9} dx$

Consider $x^2 + 2x + 9$

By comparing $\underline{x^2 + 2x}$ with $\underline{a^2 + 2ab}$, we find $a = x$; $2ab = 2x => b = 1$

$x^2 + 2x + 9 = \underline{x^2 + 2.1.x + 1^2 - 1^2 + 9}$ [add and subtract b^2 ie 1^2]

$= (x + 1)^2 + 8 = (x + 1)^2 + (\sqrt{8})^2$

$\int \frac{1}{x^2+2x+9} dx = \int \frac{1}{(x+1)^2+(\sqrt{8})^2} dx = \frac{1}{\sqrt{8}} \tan^{-1} \frac{x+1}{\sqrt{8}} + c$

3. $\int \frac{1}{3x^2+13x-10} dx$

$3x^2 + 13x - 10 \quad = 3(x^2 + \frac{13}{3}x - \frac{10}{3})$

$= 3 \left[x^2 + 2.\frac{13}{6}.x + (\frac{13}{6})^2 - (\frac{13}{6})^2 - \frac{10}{3} \right]$

$= 3 \left[\left(x + \frac{13}{6} \right)^2 - \frac{289}{36} \right] = 3[(x + \frac{13}{6})^2 - (\frac{17}{6})^2]$

$\int \frac{1}{3x^2+13x-10} dx \quad = \frac{1}{3} \int \frac{1}{(x+\frac{13}{6})^2-(\frac{17}{6})^2} dx = \frac{1}{3} \cdot \frac{1}{2(\frac{17}{6})} \log \left(\frac{x+\frac{13}{6}-\frac{17}{6}}{x+\frac{13}{6}+\frac{17}{6}} \right) + c$

$= \frac{1}{17} \log \left(\frac{6x-4}{6x+30} \right) + c = \frac{1}{17} \log \left(\frac{3x-2}{3x+15} \right) + c$

4. $\int \frac{1}{2x^2+7x+13} dx$

$2x^2 + 7x + 13 \quad = 2(x^2 + \frac{7}{2}x + \frac{13}{2})$

$= 2 \left[x^2 + 2.\frac{7}{4}.x + (\frac{7}{4})^2 - (\frac{7}{4})^2 + \frac{13}{2} \right]$

$= 2 \left[\left(x + \frac{7}{4} \right)^2 + \frac{55}{16} \right] = 2 [(x + \frac{7}{4})^2 + (\frac{\sqrt{55}}{4})^2]$

$$\int \frac{1}{2x^2+7x+13}\,dx \qquad = \frac{1}{2}\int \frac{1}{(x+\frac{7}{4})^2+(\frac{\sqrt{55}}{4})^2}\,dx$$

$$= \frac{1}{2}\cdot\frac{4}{\sqrt{55}}\tan^{-1}\frac{4\,(x+\frac{7}{4})}{\sqrt{55}}+c = \frac{2}{\sqrt{55}}\tan^{-1}\frac{4x+7}{\sqrt{55}}+c$$

5. $\int \dfrac{1}{5x^2-2x}\,dx$

$$5x^2-2x \quad = 5(x^2-\tfrac{2}{5}x) = 5(x^2-2.\tfrac{1}{5}x)$$

$$= 5\,[\,x^2-2.\tfrac{1}{5}.x+(\tfrac{1}{5})^2-(\tfrac{1}{5})^2\,]$$

$$= 5\,[(x-\tfrac{1}{5})^2-(\tfrac{1}{5})^2] = 5\,[(\,x-\tfrac{1}{5})^2-(\tfrac{1}{5})^2]$$

$$\int \frac{1}{5x^2-2x}\,dx \qquad = \frac{1}{5}\int \frac{1}{(x-\frac{1}{5})^2-(\frac{1}{5})^2}\,dx$$

$$= \frac{1}{5}\cdot\frac{5}{2}\log\left(\frac{x-\frac{1}{5}-\frac{1}{5}}{x-\frac{1}{5}+\frac{1}{5}}\right)+c = \frac{1}{2}\log\left(\frac{5x-2}{5x}\right)+c$$

6. $\int \dfrac{1}{4x^2-4x+2}\,dx$

$$4x^2-4x+2 = 4(x^2-x+\tfrac{1}{2})$$

$$= 4\,[\,x^2-2.\tfrac{1}{2}.x+(\tfrac{1}{2})^2-(\tfrac{1}{2})^2+\tfrac{1}{2}]$$

$$= 4\,[(x-\tfrac{1}{2})^2+\tfrac{1}{4}] \quad = 4\,[(\,x-\tfrac{1}{2})^2+(\tfrac{1}{2})^2]$$

$$\int \frac{1}{3x^2+13x-10}\,dx \quad = \frac{1}{4}\int \frac{1}{(x-\frac{1}{2})^2+(\frac{1}{2})^2}\,dx$$

$$= \frac{2}{4}\tan^{-1}2(x-\tfrac{1}{2})+c = \frac{1}{2}\tan^{-1}(2x-1)+c$$

Exercise 4.3.1

1. $\int \dfrac{1}{x^2+3x+1}\,dx \ = \ \ldots\ldots$

2. $\int \dfrac{1}{2x^2+8x+9}\,dx \ = \ \ldots\ldots$

3. $\int \dfrac{1}{3x-5x^2}\,dx \ = \ \ldots\ldots$

4. $\int \dfrac{1}{7+8x-x^2}\,dx \ = \ \ldots\ldots$

5. $\int \dfrac{x^3}{4+2x^4-x^8}\,dx \ = \ \ldots\ldots$

Solution 4.3.1

1. $\int \dfrac{1}{x^2+3x+1}\,dx$

$x^2 + 3x + 1 = x^2 + 2.\dfrac{3}{2}.x + (\dfrac{3}{2})^2 - (\dfrac{3}{2})^2 + 1 = (x + \dfrac{3}{2})^2 - \dfrac{5}{4}$

$x^2 + 3x + 1 = (x + \dfrac{3}{2})^2 - (\dfrac{\sqrt{5}}{2})^2$

$\int \dfrac{1}{x^2+3x+1}\,dx = \int \dfrac{1}{(x+\frac{3}{2})^2-(\frac{\sqrt{5}}{2})^2}\,dx$

$\qquad = \dfrac{1}{2(\frac{\sqrt{5}}{2})} \log\left(\dfrac{x+\frac{3}{2}-\frac{\sqrt{5}}{2}}{x+\frac{3}{2}+\frac{\sqrt{5}}{2}}\right) + c = \dfrac{1}{\sqrt{5}} \log\left(\dfrac{2x+3-\sqrt{5}}{2x+3+\sqrt{5}}\right) + c$

2. $\int \dfrac{1}{2x^2+8x+9}\,dx$

$2x^2 + 8x + 9 = 2(x^2 + 4x + \dfrac{9}{2}) = 2[\,x^2 + 2.2.x + 2^2 - 2^2 + \dfrac{9}{2}\,]$

$\qquad = 2[(x + 2)^2 + \dfrac{1}{2}] = 2[(x + 2)^2 + (\dfrac{1}{\sqrt{2}})^2]$

$\int \dfrac{1}{2x^2+8x+9}\,dx \qquad = \dfrac{1}{2}\int \dfrac{1}{(x+2)^2+(\frac{1}{\sqrt{2}})^2}\,dx$

$\qquad = \dfrac{\sqrt{2}}{2}\tan^{-1}\sqrt{2}(x + 2) + c = \dfrac{1}{\sqrt{2}}\tan^{-1}\sqrt{2}(x + 2) + c$

3. $\int \dfrac{1}{3x-5x^2}\,dx$

$3x - 5x^2 \quad = -(5x^2 - 3x) \quad = -5(x^2 - \dfrac{3}{5}x)$

$\qquad = -5(x^2 - 2.\dfrac{3}{10}x) = -5[\,x^2 - 2.\dfrac{3}{10}.x + (\dfrac{3}{10})^2 - (\dfrac{3}{10})^2\,]$

$\qquad = -5[(x - \dfrac{3}{10})^2 - (\dfrac{3}{10})^2] = 5[(\dfrac{3}{10})^2 - (x - \dfrac{3}{10})^2]$

$\int \dfrac{1}{3x-5x^2}\,dx \quad = \dfrac{1}{5}\int \dfrac{1}{(\frac{3}{10})^2-(-\frac{3}{10})^2}\,dx = \dfrac{1}{3}\log\left(\dfrac{\frac{3}{10}+(x-\frac{3}{10})}{\frac{3}{10}-(x-\frac{3}{10})}\right) + c$

$\qquad = \dfrac{1}{3}\log\left(\dfrac{3+10x-3}{3-10x+3}\right) + c = \dfrac{1}{3}\log\left(\dfrac{5x}{3-5x}\right) + c$

4. $\int \dfrac{1}{7+8x-x^2}\,dx$

$7 + 8x - x^2 \qquad = -(x^2 - 8x - 7) = -[\,x^2 - 2.4.x + 4^2 - 4^2 - 7\,]$

$$= -[(x-4)^2 - (\sqrt{23})^2] = (\sqrt{23})^2 - (x-4)^2$$

$$\int \frac{1}{7+8x-x^2}\,dx \quad = \int \frac{1}{(\sqrt{23})^2-(-4)^2}\,dx$$

$$= \frac{1}{2\sqrt{23}}\log\left(\frac{\sqrt{23}+(-4)}{\sqrt{23}-(x-4)}\right)+c = \frac{1}{2\sqrt{23}}\log\left(\frac{\sqrt{23}+x-4}{\sqrt{23}-x+4}\right)+c$$

5. $\int \dfrac{x^3}{4+2x^4-x^8}\,dx$

Let $x^4 = t \ \Rightarrow \ 4x^3dx = dt \Rightarrow x^3dx = \dfrac{1}{4}dt$

$$\int \frac{x^3}{4+2x^4-x^8}\,dx \quad = \frac{1}{4}\int \frac{1}{4+2t-t^2}\,dt$$

$$4+2t-t^2 \quad = -t^2+2t+4 = -(t^2-2t-4)$$

$$= -[t^2-2.1.t+1^2-1^2-4]$$

$$= -[(t-1)^2-(\sqrt{5})^2] = (\sqrt{5})^2-(t-1)^2$$

$$\int \frac{x^3}{4+2x^4-x^8}\,dx \quad = \frac{1}{4}\int \frac{1}{(\sqrt{5})^2-(t-1)^2}\,dx = \frac{1}{8\sqrt{5}}\log\frac{\sqrt{5}+t-1}{\sqrt{5}-t+1}+c$$

$$= \frac{1}{8\sqrt{5}}\log\frac{\sqrt{5}+x^4-1}{\sqrt{5}-x^4+1}+c$$

4.3.2 $\int \dfrac{1}{\sqrt{ax^2+bx+c}}\,dx$

Formulae required to solve this type of problems

$$\int \frac{1}{\sqrt{x^2-a^2}}\,dx \ = \ \cosh^{-1}\frac{x}{a}+c \ \text{(or)}\ \log\left(x+\sqrt{x^2-a^2}\right)+c'$$

$$\int \frac{1}{\sqrt{x^2+a^2}}\,dx \ = \ \sinh^{-1}\frac{x}{a}+c \ \text{(or)}\ \log\left(x+\sqrt{x^2+a^2}\right)+c'$$

$$\int \frac{1}{\sqrt{a^2-x^2}}\,dx \ = \ \sin^{-1}\frac{x}{a}+c$$

An Important note

As the methods involved in solving the above two type problems (4.3.1 and 4.3.2) are the same and the difference is only in the formulae applied, students can try to complete the missing steps in the following problems. Missing answers are given at the end.

Examples 4.3.2

1. $\int \dfrac{1}{\sqrt{x^2+6x+5}}\,dx$

$x^2 + 6x + 5 \qquad = \underline{x^2 + 2.3.x + 3^2 - 3^2 + 5}$ [add and subtract b^2 ie 3^2]

$\qquad\qquad\qquad\qquad = (x+3)^2 - 4 = (x+3)^2 - 2^2$

$\int \dfrac{1}{\sqrt{x^2+6x+5}}\,dx \qquad = \int \dfrac{1}{\sqrt{(x+3)^2-2^2}}\,dx$

$\qquad\qquad\qquad\qquad = \ldots\ldots\ldots\ldots\ldots\ldots\ldots$ (or) $\ldots\ldots\ldots\ldots\ldots\ldots\ldots$

2. $\int \dfrac{1}{\sqrt{x^2+2x+9}}\,dx$

$x^2 + 2x + 9 \qquad = \ldots\ldots\ldots\ldots\ldots\ldots\ldots = \ldots\ldots\ldots\ldots\ldots\ldots\ldots$

$x^2 + 2x + 9 \qquad = (x+1)^2 + (\sqrt{8})^2$

$\int \dfrac{1}{\sqrt{x^2+2x+9}}\,dx \qquad = \int \dfrac{1}{\sqrt{(x+1)^2+(\sqrt{8})^2}}\,dx$

$\qquad\qquad\qquad\qquad = \ldots\ldots\ldots\ldots\ldots\ldots\ldots$ (or) $\ldots\ldots\ldots\ldots\ldots\ldots\ldots$

3. $\int \dfrac{1}{\sqrt{3x^2+13x-10}}\,dx$

$3x^2 + 13x - 10 \qquad = \ldots\ldots\ldots\ldots\ldots\ldots\ldots = \ldots\ldots\ldots\ldots\ldots\ldots\ldots$

$\qquad\qquad\qquad\qquad = \ldots\ldots\ldots\ldots\ldots\ldots\ldots$

$3x^2 + 13x - 10 \qquad = 3\left[(x+\tfrac{13}{6})^2 - (\tfrac{17}{6})^2\right]$

$\int \dfrac{1}{\sqrt{3x^2+13x-10}}\,dx \;=\; \dfrac{1}{\sqrt{3}}\int \dfrac{1}{\sqrt{(x+\frac{13}{6})^2-(\frac{17}{6})^2}}\,dx$

$\qquad\qquad\qquad\qquad = \dfrac{1}{\sqrt{3}}\cosh^{-1}\dfrac{6x+13}{17} + c$

$\qquad\qquad\text{(or) } \dfrac{1}{\sqrt{3}}\log\left(\dfrac{6x+13}{6} + \sqrt{\left(\dfrac{6x+13}{6}\right)^2 - (\tfrac{17}{6})^2}\right) + c'$

4. $\int \dfrac{1}{\sqrt{2x^2+7x+13}}\,dx$

$2x^2 + 7x + 13 \qquad = \ldots\ldots\ldots\ldots\ldots\ldots = \ldots\ldots\ldots\ldots\ldots\ldots$

$\qquad\qquad\qquad\qquad = \ldots\ldots\ldots\ldots\ldots\ldots$

$2x^2 + 7x + 13 \qquad = 2\left[\left(x + \frac{7}{4}\right)^2 + \left(\frac{\sqrt{55}}{4}\right)^2\right]$

$\displaystyle\int \frac{1}{\sqrt{3x^2+13x-10}}\,dx \quad = \frac{1}{\sqrt{2}}\int \frac{1}{\sqrt{\left(x+\frac{7}{4}\right)^2+\left(\frac{\sqrt{55}}{4}\right)^2}}\,dx$

$\qquad\qquad = \frac{1}{\sqrt{2}}\sinh^{-1}\frac{4x+7}{\sqrt{55}} + c$

$\qquad\qquad \text{(or) } \frac{1}{\sqrt{2}}\log\left[\frac{4x+7}{4} + \sqrt{\left(\frac{4x+7}{4}\right)^2 + \frac{55}{16}}\right] + c'$

5. $\displaystyle\int \frac{1}{\sqrt{5x^2-2x}}\,dx$

$5x^2 - 2x \qquad = \ldots\ldots\ldots\ldots\ldots\ldots = \ldots\ldots\ldots\ldots\ldots\ldots$

$\qquad\qquad = \ldots\ldots\ldots\ldots\ldots\ldots$

$5x^2 - 2x \qquad = 5\left[\left(x - \frac{1}{5}\right)^2 - \left(\frac{1}{5}\right)^2\right]$

$\displaystyle\int \frac{1}{\sqrt{5x^2-2x}}\,dx \quad = \frac{1}{\sqrt{5}}\int \frac{1}{\sqrt{\left(x-\frac{1}{5}\right)^2-\left(\frac{1}{5}\right)^2}}\,dx$

$\qquad\qquad = \ldots\ldots\ldots\ldots\ldots\ldots \text{ (or) } \ldots\ldots\ldots\ldots\ldots\ldots$

6. $\displaystyle\int \frac{1}{\sqrt{4x^2-4x+2}}\,dx$

$4x^2 - 4x + 2 \qquad = \ldots\ldots\ldots\ldots\ldots\ldots = \ldots\ldots\ldots\ldots\ldots\ldots$

$\qquad\qquad = \ldots\ldots\ldots\ldots\ldots\ldots$

$\displaystyle\int \frac{1}{\sqrt{4x^2-4x+2}}\,dx \quad = \frac{1}{2}\int \frac{1}{\sqrt{\left(x+\frac{1}{2}\right)^2+\left(\frac{1}{2}\right)^2}}\,dx$

$\qquad\qquad = \frac{1}{2}\sinh^{-1}(2x+1) + c$

$\qquad\qquad \text{(or) } \frac{1}{2}\log\left[\frac{2x+1}{2} + \sqrt{\left(\frac{2x+1}{2}\right)^2 + \frac{1}{4}}\right] + c'$

Answers to examples 4.3.2

1. $\cosh^{-1}\frac{x+3}{2} + c \quad$ (or) $\log\left[(x+3) + \sqrt{(x+3)^2 - 4}\right] + c'$

2. $\sinh^{-1}\frac{x+1}{\sqrt{8}} + c \quad$ (or) $\log\left[(x+1) + \sqrt{(x+1)^2 + 8}\right] + c'$

5. $\frac{1}{\sqrt{5}} \cosh^{-1}(5x - 1) + c$ (or) $\frac{1}{\sqrt{5}} \log\left[\frac{5x-1}{5} + \sqrt{\left(\frac{5x-1}{5}\right)^2 - \left(\frac{1}{5}\right)^2}\right] + c'$

Exercise 4.3.2

1. $\int \frac{1}{\sqrt{x^2+3x+1}} dx$ = ………

2. $\int \frac{1}{\sqrt{2x^2+8x+9}} dx$ = ………

3. $\int \frac{1}{\sqrt{3x-5x^2}} dx$ = ………

4. $\int \frac{1}{\sqrt{7+8x-x^2}} dx$ = ………

5. $\int \frac{x^3}{\sqrt{4+2x^4-x^8}} dx =$ ………

Solution 4.3.2

1. $\int \frac{1}{\sqrt{x^2+3x+1}} dx$

$x^2 + 3x + 1$ = …………………………

$x^2 + 3x + 1$ = $\left(x + \frac{3}{2}\right)^2 - \left(\frac{\sqrt{5}}{2}\right)^2$

$\int \frac{1}{\sqrt{x^2+2x+1}} dx$ = $\int \frac{1}{\sqrt{\left(x+\frac{3}{2}\right)^2 - \left(\frac{\sqrt{5}}{2}\right)^2}} dx$

= ……………………… (or) …………………………

2. $\int \frac{1}{\sqrt{2x^2+8x+9}} dx$

$2x^2 + 8x + 9$ = …………………………= …………………………

=………………………

$2x^2 + 8x + 9$ = $2\left[(x + 2)^2 + \left(\frac{1}{\sqrt{2}}\right)^2\right]$

$\int \frac{1}{\sqrt{2x^2+8x+9}} dx$ = $\frac{1}{\sqrt{2}} \int$ …… … … … … … … … dx

= ……………………… (or) ………………………

3. $\int \frac{1}{\sqrt{3x-5x^2}} dx$

$3x - 5x^2$ = $-(5x^2 - 3x)$

= ………………………… = …………………………

= …………………………=…………………………

$3x - 5x^2 \qquad = 5[\left(\frac{3}{10}\right)^2 - (x - \frac{3}{10})^2]$

$\int \frac{1}{\sqrt{3x - 5x^2}} dx \qquad = \frac{1}{\sqrt{5}} \int \frac{1}{\sqrt{\left(\frac{3}{10}\right)^2 - (-\frac{3}{10})^2}} dx$

$\qquad\qquad\qquad\qquad = \dots\dots\dots\dots\dots\dots$

4. $\int \frac{1}{\sqrt{7 + 8x - x^2}} dx$

$7 + 8x - x^2 \qquad = \dots\dots\dots\dots\dots\dots\dots\dots = \dots\dots\dots\dots\dots\dots\dots$

$\qquad\qquad\qquad\qquad = \dots\dots\dots\dots\dots\dots$

$\int \frac{1}{\sqrt{7 + 8x - x^2}} dx \qquad = \int \frac{1}{\sqrt{(\sqrt{23})^2 - (-4)^2}} dx$

$\qquad\qquad\qquad\qquad = \dots\dots\dots\dots\dots\dots$

5. $\int \frac{x^3}{\sqrt{4 + 2x^4 - x^8}} dx$

Let $x^4 = t \implies 4x^3 dx = dt \implies x^3 dx = \frac{1}{4} dt$

$\int \frac{x^3}{\sqrt{4 + 2x^4 - x^8}} dx \qquad = \frac{1}{4} \int \frac{1}{\sqrt{4 + 2t - t^2}} dt$

$4 + 2t - t^2 \qquad = -(t^2 - 2t - 4) = -[t - 2.1.t + 1^2 - 1^2 - 4]$

$\qquad\qquad\qquad\qquad = -[(t - 1)^2 - (\sqrt{5})^2] = (\sqrt{5})^2 - (t - 1)^2$

$\int \frac{x^3}{\sqrt{4 + 2x^4 - x^8}} dx \qquad = \frac{1}{4} \int \frac{1}{\sqrt{(\sqrt{5})^2 - (t-1)^2}} dx = \frac{1}{4} \sin^{-1} \frac{t-1}{\sqrt{5}} + c$

$\qquad\qquad\qquad\qquad = \frac{1}{4} \sin^{-1} \frac{x^4 - 1}{\sqrt{5}} + c$

Answers to exercise 4.3.2

1. $\cosh^{-1}(\frac{2x+3}{\sqrt{5}}) + c$ (or) $\log [\frac{2x+3}{2} + \sqrt{(\frac{2x+3}{2})^2 - (\frac{\sqrt{5}}{2})^2}] + c'$

2. $\frac{1}{\sqrt{2}} \sinh^{-1} \sqrt{2}(x + 2) + c$ (or) $\frac{1}{\sqrt{2}} \log [(x + 2) + \sqrt{(x + 2)^2 + \frac{1}{2}}] + c'$

3. $\frac{1}{\sqrt{5}} \sin^{-1} \frac{10x-3}{3} + c$ \qquad\qquad 4. $\sin^{-1}(\frac{x-4}{\sqrt{23}}) + c$

Unit 4.1 - 4.3   Try to recollect the following

Integration

4.1	$\int \frac{1}{\sqrt{x^2-a^2}} dx = \cosh^{-1}\frac{x}{a} + c$ (or) $\log(x + \sqrt{x^2 - a^2}) + c'$
	$\int \frac{1}{\sqrt{x^2+a^2}} dx = \sinh^{-1}\frac{x}{a} + c$ (or) $\log(x + \sqrt{x^2 + a^2}) + c'$
	$\int \frac{1}{\sqrt{a^2-x^2}} dx = \sin^{-1}\frac{x}{a} + c$
4.2	$\int \frac{1}{x^2-a^2} dx = \frac{1}{2a}\log\left(\frac{x-a}{x+a}\right) + c$; $\int \frac{1}{a^2-x^2} dx = \frac{1}{2a}\log\left(\frac{a+x}{a-x}\right) + c$
	$\int \frac{1}{x^2+a^2} dx = \frac{1}{a}\tan^{-1}\frac{x}{a} + c$

4.3.1		4.3.2	
Integrals of the form $\int \frac{1}{ax^2+bx+c} dx$		Integrals of the form $\int \frac{1}{\sqrt{ax^2+bx+c}} dx$	

Remember the following substitution

For   i. $x^2 - a^2$, substitute $x = a\sec\theta$

ii. $x^2 + a^2$, substitute $x = a\tan\theta$

iii. $a^2 - x^2$, substitute $x = a\sin\theta$

Self Evaluation Test 4 :

Match the following I

	Column I		Column II
1	$\int \frac{1}{\sqrt{a^2-x^2}}\ dx$	I	$\frac{1}{2a}\log(\frac{x-a}{x+a}) + c$
2	$\int \frac{1}{a^2-x^2}\ dx$	II	$\cosh^{-1}\frac{x}{a} + c$
3	$\int \frac{1}{\sqrt{x^2-a^2}}\ dx$	III	$\sinh^{-1}\frac{x}{a} + c$
4	$\int \frac{1}{x^2+a^2}\ dx$	IV	$\sin^{-1}\frac{x}{a} + c$
5	$\int \frac{1}{x^2-a^2}\ dx$	V	$\frac{1}{a}\ \tan^{-1}\frac{x}{a} + c$
6	$\int \frac{1}{\sqrt{x^2+a^2}}\ dx$	VI	$\log(x + \sqrt{x^2 + a^2}) + c'$
		VII	$\frac{1}{2a}\log(\frac{a+x}{a-x}) + c$
		VIII	$\log(x + \sqrt{x^2 - a^2}) + c'$

Answers : 1. IV   2. VII   3. II, VIII   4. V   5. I   6. III, VI

Match the following II

	Column I		Column II
1	$2x^2 + 6x + 5$	I	$(\sqrt{23})^2 - (x-4)^2$
2	$x^2 + 3x + 8$	II	$2[(x+2)^2 + (\frac{1}{\sqrt{2}})^2]$
3	$x^2 + x + 1$	III	$(x+\frac{1}{2})^2 + (\frac{\sqrt{3}}{2})^2$
4	$7 + 8x - x^2$	IV	$5[(\frac{3}{10})^2 - (x-\frac{3}{10})^2]$
5	$2x^2 + 8x + 9$	V	$(x+\frac{3}{2})^2 + (\frac{\sqrt{23}}{2})^2$
6	$3x - 5x^2$	VI	$2[(x+\frac{3}{2})^2 + (\frac{1}{2})^2]$

Answers : 1. VI   2. V   3. III   4. I   5. II   6. IV

4.4 Integrals of the form $\int \frac{px+q}{ax^2+bx+c} dx$ and $\int \frac{px+q}{\sqrt{ax^2+bx+c}} dx$

4.4.1 Integrals of the form $\int \frac{px+q}{ax^2+bx+c} dx$

Method : Find two real numbers A, B using

$$px + q = A \frac{d}{dx}(ax^2 + bx + c) + B = A(2ax + b) + B$$

by equating the coefficients of x and the constant term.

Examples

1. $\int \frac{x+2}{2x^2+6x+5} dx$

Step 1. To find the values of A and B

$$x + 2 = A \frac{d}{dx}(2x^2 + 6x + 5) + B = A(4x + 6) + B \quad \text{-------------- (i)}$$

Equating coeff. of x both sides we get, $\quad 4A = 1 \implies A = \frac{1}{4}$

Equating the constants , we get, $6A + B = 2 \implies 6.\frac{1}{4} + B = 2$ ie. $B = \frac{1}{2}$

(1) becomes $\quad x + 2 = \frac{1}{4}(4x + 6) + \frac{1}{2}$

$$\int \frac{x+2}{2x^2+6x+5} dx = \frac{1}{4} \int \frac{4x+6}{2x^2+6x+5} dx + \frac{1}{2} \int \frac{1}{2x^2+6x+5} dx = \frac{1}{4}I + \frac{1}{2}I' \quad \text{...... (ii)}$$

where $I = \int \frac{4x+6}{2x^2+6x+5} dx \quad$ and $\quad I' = \int \frac{1}{2x^2+6x+5} dx$

Step 2 To evaluate the integrals I and I$'$

$$I = \int \frac{4x+6}{2x^2+6x+5} dx = \log(2x^2 + 6x + 5) + c' \quad [\int \frac{f'(x)}{f(x)} dx = \log[f(x)] + c]$$

$$I' = \int \frac{1}{2x^2+6x+5} dx$$

$$2x^2 + 6x + 5 = 2(x^2 + 3x + \frac{5}{2}) = 2[x^2 + 2.\frac{3}{2}x + (\frac{3}{2})^2 - (\frac{3}{2})^2 + \frac{5}{2}]$$

$$= 2[(x + \frac{3}{2})^2 - \frac{9}{4} + \frac{5}{2}] = 2[(x + \frac{3}{2})^2 + (\frac{1}{2})^2]$$

$$I' = \int \frac{1}{2x^2+6x+5} dx = \frac{1}{2} \int \frac{1}{(x+\frac{3}{2})^2 + (\frac{1}{2})^2} dx$$

$$= \frac{1}{2}.2 \tan^{-1} 2(x + \frac{3}{2}) + c'' = \tan^{-1}(2x + 3) + c''$$

Step 3 Substituting the values of I and I' in (ii), we get

$$\int \frac{x+2}{2x^2+6x+5} dx \quad = \quad \frac{1}{4}I + \frac{1}{2}I'$$

$$=> \int \frac{x+2}{2x^2+6x+5} dx \quad = \frac{1}{4}\log(2x^2+6x+5) + \frac{1}{4}c' + \frac{1}{2}\tan^{-1}(2x+3) + \frac{1}{2}c''$$

$$= \frac{1}{4}\log(2x^2+6x+5) + \frac{1}{2}\tan^{-1}(2x+3) + c \, , c = \frac{1}{4}c' + \frac{1}{2}c''$$

2. $\int \frac{4x-3}{x^2+3x+8} dx$

Step 1. To find the values of A and B

$$4x - 3 = A\frac{d}{dx}(x^2+3x+8) + B = A(2x+3) + B \qquad \text{------------- (i)}$$

Equating coeff. of x both sides we get, $2A = 4 \implies A = 2$

Equating the constants, we get $3A + B = -3 \implies 6 + B = -3$, ie. $B = -9$

(1) becomes $4x - 3 = 2(2x+3) - 9$

$$\int \frac{4x-3}{x^2+3x+8} dx = 2\int \frac{2x+3}{x^2+3x+8} dx - 9\int \frac{1}{x^2+3x+8} dx = 2I - 9I' \quad \text{...... (ii)}$$

where $I = \int \frac{2x+3}{x^2+3x+8} dx$ and $I' = \int \frac{1}{x^2+3x+8} dx$

Step 2 To evaluate the integrals I and I'

$$I = \int \frac{2x+3}{x^2+3x+8} dx = \log(x^2+3x+8) + c'$$

$$I' = \int \frac{1}{x^2+3x+8} dx$$

$$x^2 + 3x + 8 = x^2 + 2.\frac{3}{2}x + (\frac{3}{2})^2 - (\frac{3}{2})^2 + 8$$

$$= (x+\frac{3}{2})^2 - \frac{9}{4} + 8 = (x+\frac{3}{2})^2 + (\frac{\sqrt{23}}{2})^2$$

$$I' = \int \frac{1}{x^2+3x+8} dx = \int \frac{1}{(x+\frac{3}{2})^2 + (\frac{\sqrt{23}}{2})^2} dx$$

$$= \frac{2}{\sqrt{23}}\tan^{-1}[\frac{2}{\sqrt{23}}(x+\frac{3}{2})] + c'' = \frac{2}{\sqrt{23}}\tan^{-1}(\frac{2x+3}{\sqrt{23}}) + c''$$

Step 3 Substituting the values of I and I' in (ii), we get

$$\int \frac{4x-3}{x^2+3x+8} dx = 2I - 9I'$$

$$=> \int \frac{4x-3}{x^2+3x+8} dx = 2\log(x^2+3x+8) + 2c' - \frac{18}{\sqrt{23}}\tan^{-1}\left(\frac{2x+3}{\sqrt{23}}\right) - 9c''$$

$$= 2\log(x^2+3x+8) - \frac{18}{\sqrt{23}}\tan^{-1}\left(\frac{2x+3}{\sqrt{23}}\right) + c, \quad c = 2c' - 9c''$$

3. $\int \frac{5x+1}{x^2-2x-35} dx$

Step 1. To find the values of A and B

$$5x + 1 = A \frac{d}{dx}(x^2 - 2x - 35) + B = A(2x - 2) + B \quad \text{------------- (i)}$$

Equating coeff. of x both sides we get, $2A = 5 \implies A = \frac{5}{2}$

Equating the constants, we get, $-2A + B = 1 \implies -5 + B = 1$, ie. $B = 6$

(1) becomes $5x + 1 = \frac{5}{2}(2x - 2) + 6$

$$\int \frac{5x+1}{x^2-2x-35} dx = \frac{5}{2}\int \frac{2x-2}{x^2-2x-35} dx + 6\int \frac{1}{x^2-2x-35} dx = \frac{5}{2}I + 6I' \quad \text{...... (ii)}$$

where $I = \int \frac{2x-2}{x^2-2x-35} dx$ and $I' = \int \frac{1}{x^2-2x-35} dx$

Step 2 To evaluate the integrals I and I'

$$I = \int \frac{2x-2}{x^2-2x-35} dx = \log(x^2 - 2x - 35) + c'$$

$$I' = \int \frac{1}{x^2-2x-35} dx$$

$$x^2 - 2x - 35 = x^2 - 2.1x + 1^2 - 1^2 - 35 = (x - 1)^2 - 6^2$$

$$I' = \int \frac{1}{(x-1)^2 - 6^2} dx = \frac{1}{12}\log\left(\frac{x-1-6}{x-1+6}\right) + c'' = \frac{1}{12}\log\left(\frac{x-7}{x+5}\right) + c''$$

Step 3 Substituting the values of I and I' in (ii), we get

$$\int \frac{5x+1}{x^2-2x-35} dx = \frac{5}{2}I + 6I'$$

$$=> \int \frac{5x+1}{x^2-2x-35} dx = \frac{5}{2}\log(x^2 - 2x - 35) + \frac{5c'}{2} + \frac{1}{2}\log\left(\frac{x-7}{x+5}\right) + 6c''$$

$$= \frac{5}{2}\log(x^2 - 2x - 35) + \frac{1}{2}\log\left(\frac{x-7}{x+5}\right) + c, \quad c = \frac{5c'}{2} + 6c''$$

Exercise 4.4.1

1. $\int \frac{5x-2}{3x^2+2x+1} dx = $ ...... 2. $\int \frac{2x+3}{x^2+x+1} dx = $ ...... 3. $\int \frac{x+3}{x^2-2x-5} dx = $ ......

Solution 4.4.1

1. $\int \frac{5x-2}{3x^2+2x+1} dx$

Step 1. To find the values of A and B

$5x - 2 = A \frac{d}{dx}(3x^2 + 2x + 1) + B = A(6x + 2) + B$ -------------- (i)

Equating coeff. of x both sides we get, $6A = 5 \implies A = \frac{5}{6}$

Equating the constants , we get, $2A + B = -2 \implies 2.\frac{5}{6} + B = -2$, ie. $B = -\frac{11}{3}$

(1) becomes $5x - 2 = \frac{5}{6}(6x + 2) - \frac{11}{3}$

$\int \frac{5x-2}{3x^2+2x+1} dx = \frac{5}{6}\int \frac{6x+2}{3x^2+2x+1} dx - \frac{11}{3}\int \frac{1}{3x^2+2x+1} dx = \frac{5}{6}I - \frac{11}{3}I'$ ...... (2)

Where $I = \int \frac{6x+2}{3x^2+2x+1} dx$ and $I' = \int \frac{1}{3x^2+2x+1} dx$

Step 2 To evaluate the integrals I and I'

$I = \int \frac{6x+2}{3x^2+2x+1} dx = \log(3x^2 + 2x + 1) + c'$

$I' = \int \frac{1}{3x^2+2x+1} dx$

$3x^2 + 2x + 1 = 3(x^2 + \frac{2}{3}x + \frac{1}{3}) = 3[x^2 + 2.\frac{1}{3}x + (\frac{1}{3})^2 - (\frac{1}{3})^2 + \frac{1}{3}]$

$= 3[(x+\frac{1}{3})^2 - \frac{1}{9} + \frac{1}{3}] = 3[(x+\frac{1}{3})^2 + (\frac{\sqrt{2}}{3})^2]$

$I' = \int \frac{1}{3x^2+2x+1} dx = \frac{1}{3}\int \frac{1}{(x+\frac{1}{3})^2 + (\frac{\sqrt{2}}{3})^2} dx = \frac{1}{3}.\frac{3}{\sqrt{2}}\tan^{-1}[\frac{3}{\sqrt{2}}(x+\frac{1}{3})] + c''$

$= \frac{1}{\sqrt{2}}\tan^{-1}\frac{(3x+1)}{\sqrt{2}} + c''$

Step 3 Substituting the values of I and I' in (ii), we get

$\int \frac{5x-2}{3x^2+2x+1} dx = \frac{5}{6}I - \frac{11}{3}I'$

$\implies \int \frac{5x-2}{3x^2+2x+1} dx = \frac{5}{6}\log(3x^2 + 2x + 1) + \frac{5}{6}c' - \frac{11}{3}.\frac{1}{\sqrt{2}}\tan^{-1}\frac{(3x+1)}{\sqrt{2}} - \frac{11}{3}c''$

$= \frac{5}{6}\log(3x^2 + 2x + 1) - \frac{11}{3\sqrt{2}}\tan^{-1}\frac{(3x+1)}{\sqrt{2}} + c, \ c = \frac{5}{6}c' - \frac{11}{3}c''$

2. $\int \frac{2x+3}{x^2+x+1} dx$

Step 1. To find the values of A and B

$2x + 3 = A \frac{d}{dx}(x^2 + x + 1) + B = A(2x + 1) + B$ -------------- (i)

Equating coeff. of x both sides we get, $2A = 2 \implies A = 1$

Equating the constants we get, $A + B = 3 \implies 1 + B = 3$, ie. $B = 2$

(1) becomes $2x + 3 = (2x + 1) + 2$

$\int \frac{2x+3}{x^2+x+1} dx = \int \frac{2x+1}{x^2+x+1} dx + 2 \int \frac{1}{x^2+x+1} dx = I + 2I'$ ...... (ii)

where $I = \int \frac{2x+1}{x^2+x+1} dx$ and $I' = \int \frac{1}{x^2+x+1} dx$

Step 2 To evaluate the integrals I and I'

$I = \int \frac{2x+1}{x^2+x+1} dx = \log(x^2 + x + 1) + c'$

$I' = \int \frac{1}{x^2+x+1} dx$

$x^2 + x + 1 \qquad = x^2 + 2.\frac{1}{2}x + (\frac{1}{2})^2 - (\frac{1}{2})^2 + 1$

$\qquad\qquad = (x + \frac{1}{2})^2 - \frac{1}{4} + 1 = (x + \frac{1}{2})^2 + (\frac{\sqrt{3}}{2})^2$

$I' = \int \frac{1}{x^2+x+1} dx \qquad = \int \frac{1}{(x+\frac{1}{2})^2 + (\frac{\sqrt{3}}{2})^2} dx$

$\qquad\qquad = \frac{2}{\sqrt{3}} \tan^{-1} \frac{2}{\sqrt{3}}(x + \frac{1}{2}) + c'' = \frac{2}{\sqrt{3}} \tan^{-1}(\frac{2x+1}{\sqrt{3}}) + c''$

Step 3 Substituting the values of I and I' in (2), we get

$\int \frac{2x+3}{x^2+x+1} dx \qquad = I + 2I'$

$\implies \int \frac{2x+3}{x^2+x+1} dx \qquad = \log(x^2 + x + 1) + c' + 2.\frac{2}{\sqrt{3}} \tan^{-1}(\frac{2x+1}{\sqrt{3}}) + 2c''$

$\qquad\qquad = \log(x^2 + x + 1) + \frac{4}{\sqrt{3}} \tan^{-1}(\frac{2x+1}{\sqrt{3}}) + c, \ c = c' + 2c''$

3. $\int \frac{x+3}{x^2-2x-5} dx$

Step 1. To find the values of A and B

$$x + 3 = A \frac{d}{dx}(x^2 - 2x - 5) + B = A(2x - 2) + B \qquad \text{------------- (i)}$$

Equating coeff. of x both sides we get, $2A = 1 \implies A = \frac{1}{2}$

Equating the constants we get, $-2A + B = 3 \implies -1 + B = 3$, ie. $B = 4$

(1) becomes $x + 3 = \frac{1}{2}(2x - 2) + 4$

$$\int \frac{x+3}{x^2-2x-5}\,dx = \frac{1}{2}\int \frac{2x-2}{x^2-2x-5}\,dx + 4\int \frac{1}{x^2-2x-5}\,dx = \frac{1}{2}I + 4\,I' \quad \ldots\ldots \text{(ii)}$$

where $I = \int \frac{2x-2}{x^2-2x-5}\,dx$ and $I' = \int \frac{1}{x^2-2x-5}\,dx$

Step 2 To evaluate the integrals I and I'

$$I = \int \frac{2x-2}{x^2-2x-5}\,dx = \log(x^2 - 2x - 5) + c'$$

$$I' = \int \frac{1}{x^2-2x-5}\,dx$$

$$x^2 - 2x - 5 = x^2 - 2.1x + 1^2 - 1^2 - 5 = (x-1)^2 - (\sqrt{6})^2$$

$$I' = \int \frac{1}{(x-1)^2-(\sqrt{6})^2}\,dx = \frac{1}{2\sqrt{6}}\log\left(\frac{x-1-\sqrt{6}}{x-1+\sqrt{6}}\right) + c''$$

Step 3 Substituting the values of I and I' in (2), we get

$$\int \frac{x+3}{x^2-2x-5}\,dx = \frac{1}{2}I + 4\,I'$$

$$\implies \int \frac{x+3}{x^2-2x-5}\,dx = \frac{1}{2}\log(x^2 - 2x - 5) + \frac{c'}{2} + \frac{4}{2\sqrt{6}}\log\left(\frac{x-1-\sqrt{6}}{x-1+\sqrt{6}}\right) + 4c''$$

$$= \frac{1}{2}\log(x^2 - 2x - 5) + \frac{2}{\sqrt{6}}\log\left(\frac{x-1-\sqrt{6}}{x-1+\sqrt{6}}\right) + c, \; c = \frac{c'}{2} + 4c''$$

4.4.2 Integrals of the form $\int \frac{px+q}{\sqrt{ax^2+bx+c}}\,dx$

Method is exactly similar to type 4.4.1 but there is a **difference**.

Find two real numbers A, B using

$$px + q = A \frac{d}{dx}(\textbf{\textit{expression inside the square root}}) + B$$

$$px + q = A \frac{d}{dx}(ax^2 + bx + c) + B = A(2ax + b) + B$$

by equating the coefficients of x and the constant term.

Examples 4.4.2

1. $\int \dfrac{x+2}{\sqrt{2x^2+6x+5}}\,dx$

Step 1. To find the values of A and B

$$x + 2 = A\,\frac{d}{dx}(2x^2 + 6x + 5) + B = A\,(4x + 6) + B \quad\text{-------------- (i)}$$

Equating coeff. of x both sides we get, $4A = 1 \implies A = \dfrac{1}{4}$

Equating the constants we get, $6A + B = 2 \implies 6.\dfrac{1}{4} + B = 2$, ie. $B = \dfrac{1}{2}$

(1) becomes $x + 2 = \dfrac{1}{4}(4x + 6) + \dfrac{1}{2}$

$$\int \frac{x+2}{\sqrt{2x^2+6x+5}}\,dx = \frac{1}{4}\int \frac{4x+6}{\sqrt{2x^2+6x+5}}\,dx + \frac{1}{2}\int \frac{1}{\sqrt{2x^2+6x+5}}\,dx = \frac{1}{4}I + \frac{1}{2}I' \dots\dots (2)$$

where $I = \int \dfrac{4x+6}{\sqrt{2x^2+6x+5}}\,dx$ and $I' = \int \dfrac{1}{\sqrt{2x^2+6x+5}}\,dx$

Step 2 To evaluate the integrals I and I'

$$I = \int \frac{4x+6}{\sqrt{2x^2+6x+5}}\,dx = 2\sqrt{2x^2+6x+5} + c' \qquad [\int \frac{f'(x)}{\sqrt{f(x)}}\,dx = 2\sqrt{f(x)} + c\,]$$

$$I' = \int \frac{1}{\sqrt{2x^2+6x+5}}\,dx$$

$$2x^2 + 6x + 5 = 2\left(x^2 + 3x + \frac{5}{2}\right) = 2\left[x^2 + 2.\frac{3}{2}x + \left(\frac{3}{2}\right)^2 - \left(\frac{3}{2}\right)^2 + \frac{5}{2}\right]$$

$$= 2\left[\left(x + \frac{3}{2}\right)^2 - \frac{9}{4} + \frac{5}{2}\right] = 2\left[\left(x + \frac{3}{2}\right)^2 + \left(\frac{1}{2}\right)^2\right]$$

$$I' = \int \frac{1}{\sqrt{2x^2+6x+5}}\,dx = \frac{1}{\sqrt{2}}\int \frac{1}{\sqrt{\left(x+\frac{3}{2}\right)^2 + \left(\frac{1}{2}\right)^2}}\,dx$$

$$= \frac{1}{\sqrt{2}}\sinh^{-1} 2\left(x + \frac{3}{2}\right) + c'' = \frac{1}{\sqrt{2}}\sinh^{-1}(2x + 3) + c''$$

Step 3 Substituting the values of I and I' in (2), we get

$$\int \frac{x+2}{\sqrt{2x^2+6x+5}}\,dx = \frac{1}{4}I + \frac{1}{2}I'$$

$$\int \frac{x+2}{\sqrt{2x^2+6x+5}}\,dx = \frac{1}{4}.2\sqrt{2x^2+6x+5} + \frac{1}{4}c' + \frac{1}{2\sqrt{2}}\sinh^{-1}(2x+3) + \frac{1}{2}c''$$

$$= \frac{1}{2}\sqrt{2x^2+6x+5} + \frac{1}{2\sqrt{2}}\sinh^{-1}(2x+3) + c,\ c = \frac{1}{4}c' + \frac{1}{2}c''$$

2. $\int \frac{4x-3}{\sqrt{x^2+3x+8}} dx$

Step 1. To find the values of A and B

$4x - 3 = A \frac{d}{dx}(x^2 + 3x + 8) + B = A(2x+3) + B$ ------------ (i)

Equating coeff. of x both sides we get, $2A = 4 \Rightarrow A = 2$

Equating the constants we get, $3A + B = -3 \Rightarrow 6 + B = -3$, ie. $B = -9$

(1) becomes $4x - 3 = 2(2x+3) - 9$

$\int \frac{4x-3}{\sqrt{x^2+3x+8}} dx = 2\int \frac{2x+3}{\sqrt{x^2+3x+8}} dx - 9\int \frac{1}{\sqrt{x^2+3x+8}} dx = 2I - 9I'$ ...... (ii)

Where $I = \int \frac{2x+3}{\sqrt{x^2+3x+8}} dx$ and $I' = \int \frac{1}{\sqrt{x^2+3x+8}} dx$

Step 2 To evaluate the integrals I and I′

$I = \int \frac{2x+3}{\sqrt{x^2+3x+8}} dx = 2\sqrt{x^2 + 3x + 8} + c'$ why?

Note : In this type problems, the integral I can be easily evaluated using

the known formula $\int \frac{f'(x)}{\sqrt{f(x)}} dx = 2\sqrt{f(x)} + c$ [**Note the difference**]

$I' = \int \frac{1}{\sqrt{x^2+3x+8}} dx$

$x^2 + 3x + 8 = x^2 + 2.\frac{3}{2}x + (\frac{3}{2})^2 - (\frac{3}{2})^2 + 8$

$\qquad = (x + \frac{3}{2})^2 - \frac{9}{4} + 8 = (x + \frac{3}{2})^2 + (\frac{\sqrt{23}}{2})^2$

$I' = \int \frac{1}{\sqrt{x^2+3x+8}} dx = \int \frac{1}{\sqrt{(x+\frac{3}{2})^2 + (\frac{\sqrt{23}}{2})^2}} dx$

$\qquad = \sinh^{-1}[\frac{2}{\sqrt{23}}(x + \frac{3}{2})] + c'' = \sinh^{-1}(\frac{2x+3}{\sqrt{23}}) + c''$

Step 3 Substituting the values of I and I' in (2), we get

$\int \frac{4x-3}{\sqrt{x^2+3x+8}} dx = 2I - 9I'$

$\Rightarrow \int \frac{4x-3}{\sqrt{x^2+3x+8}} dx = 4\sqrt{x^2 + 3x + 8} + 2c' - 9\sinh^{-1}(\frac{2x+3}{\sqrt{23}}) - 9c''$

$\qquad = 4\sqrt{x^2 + 3x + 8} - 9\sinh^{-1}(\frac{2x+3}{\sqrt{23}}) + c, \ c = 2c' - 9c''$

3. $\int \frac{5x+1}{\sqrt{x^2-2x-35}} dx$

Step 1. To find the values of A and B

$5x + 1 = A \frac{d}{dx}(x^2 - 2x - 35) + B = A(2x - 2) + B$ -------------- (i)

Equating coeff. of x both sides we get, $2A = 5 \Rightarrow A = \frac{5}{2}$

Equating the constants we get, $-2A + B = 1 \Rightarrow -5 + B = 1$, ie. $B = 6$

(1) becomes $5x + 1 = \frac{5}{2}(2x - 2) + 6$

$\int \frac{5x+1}{\sqrt{x^2-2x-35}} dx = \frac{5}{2}\int \frac{2x-2}{\sqrt{x^2-2x-35}} dx + 6\int \frac{1}{\sqrt{x^2-2x-35}} dx = \frac{5}{2}I + 6I'$ ...... (ii)

where $I = \int \frac{2x-2}{\sqrt{x^2-2x-35}} dx$ and $I' = \int \frac{1}{\sqrt{x^2-2x-35}} dx$

Step 2 To evaluate the integrals I and I'

$I = \int \frac{2x-2}{\sqrt{x^2-2x-35}} dx = 2\sqrt{x^2 - 2x - 35} + c'$

$I' = \int \frac{1}{\sqrt{x^2-2x-35}} dx$

$x^2 - 2x - 35 = x^2 - 2.1x + 1^2 - 1^2 - 35 = (x - 1)^2 - 6^2$

$I' = \int \frac{1}{\sqrt{(x-1)^2 - 6^2}} dx$

$\quad = \log\left(x - 1 + \sqrt{(x-1)^2 - 6^2}\right) + c''$ (or) $\cosh^{-1}\frac{x-1}{6} + c''$

$\quad = \log\left(x - 1 + \sqrt{x^2 - 2x - 35}\right) + c''$ (or) $\cosh^{-1}\frac{x-1}{6} + c''$

Step 3 Substituting the values of I and I' in (2), we get

$\int \frac{5x+1}{\sqrt{x^2-2x-35}} dx = \frac{5}{2}I + 6I'$

$\Rightarrow \int \frac{5x+1}{\sqrt{x^2-2x-35}} dx = \frac{5}{2}.2\sqrt{x^2 - 2x - 35} + c' + 6I'$

$\quad = 5\sqrt{x^2 - 2x - 35} + c' + 6I'$

where $I' = \log\left(x - 1 + \sqrt{x^2 - 2x - 35}\right) + c''$ (or) $\cosh^{-1}\frac{x-1}{6} + c''$

Exercise 4.4.2

1. $\int \dfrac{5x-2}{\sqrt{3x^2+2x+1}}\,dx \quad = \ \ldots\ldots$

2. $\int \dfrac{2x+3}{\sqrt{x^2+x+1}}\,dx \quad = \ \ldots\ldots$

3. $\int \dfrac{x+3}{\sqrt{x^2-2x-5}}\,dx \quad = \ \ldots\ldots$

Solution 4.4.2

1. $\int \dfrac{5x-2}{\sqrt{3x^2+2x+1}}\,dx$

Step 1. To find the values of A and B

$$5x - 2 \ = \ A\,\frac{d}{dx}(3x^2 + 2x + 1) + B \ = \ A\,(6x+2) + B \quad \text{------------ (i)}$$

Equating coeff. of x both sides we get, $\quad \ldots\ldots A = \ldots\ldots$ (Try to complete)

Equating the constants , we get $\quad \ldots\ldots B = \ldots\ldots$, ie. $B = -2 - \dfrac{5}{3} = -\dfrac{11}{3}$

(1) becomes $\quad 5x - 2 = \dfrac{5}{6}(6x+2) - \dfrac{11}{3}$

$$\int \frac{5x-2}{\sqrt{3x^2+2x+1}}\,dx = \frac{5}{6}\int \frac{6x+2}{\sqrt{3x^2+2x+1}}\,dx - \frac{11}{3}\int \frac{1}{\sqrt{3x^2+2x+1}}\,dx = \frac{5}{6}I - \frac{11}{3}I' \ldots\ldots \text{ (ii)}$$

where $I = \int \dfrac{6x+2}{\sqrt{3x^2+2x+1}}\,dx$ and $I' = \int \dfrac{1}{\sqrt{3x^2+2x+1}}\,dx$

Step 2 To evaluate the integrals I and I'

$$I = \int \frac{6x+2}{\sqrt{3x^2+2x+1}}\,dx = 2\sqrt{3x^2+2x+1} + c'$$

$$I' = \int \frac{1}{\sqrt{3x^2+2x+1}}\,dx$$

$$3x^2 + 2x + 1 = 3\left(x^2 + \frac{2}{3}x + \frac{1}{3}\right) = 3\left[x^2 + 2.\frac{1}{3}x + \left(\frac{1}{3}\right)^2 - \left(\frac{1}{3}\right)^2 + \frac{1}{3}\right]$$

$$= 3\left[\left(x + \frac{1}{3}\right)^2 - \frac{1}{9} + \frac{1}{3}\right] = 3\left[\left(x + \frac{1}{3}\right)^2 + \left(\frac{\sqrt{2}}{3}\right)^2\right]$$

$$I' = \frac{1}{\sqrt{3}}\int \frac{1}{\sqrt{3x^2+2x+1}}\,dx = \frac{1}{\sqrt{3}}\int \frac{1}{\sqrt{\left(x+\frac{3}{2}\right)^2 + \left(\frac{\sqrt{2}}{3}\right)^2}}\,dx$$

$$= \frac{1}{\sqrt{3}}\sinh^{-1}\left[\frac{3}{\sqrt{2}}\left(x + \frac{3}{2}\right)\right] + c'' = \frac{1}{\sqrt{3}}\sinh^{-1}\frac{3(2x+3)}{2\sqrt{2}} + c''$$

Step 3 Substituting the values of I and I' in (2), we get

$$\int \frac{5x-2}{\sqrt{3x^2+2x+1}}\,dx \quad = \frac{5}{6}I - \frac{11}{3}I'$$

$$\Rightarrow \int \frac{5x-2}{\sqrt{3x^2+2x+1}}\,dx = \frac{5}{6}.2\sqrt{3x^2+2x+1} + \frac{5}{6}c' - \frac{11}{3\sqrt{3}}\sinh^{-1}\frac{3(2x+3)}{2\sqrt{2}} - \frac{11}{3}c''$$

$$= \frac{5}{3}\sqrt{3x^2+2x+1} - \frac{11}{3\sqrt{3}}\sinh^{-1}\frac{3(2x+3)}{2\sqrt{2}} + c$$

$$\text{where } c = \frac{5}{6}c' - \frac{11}{3}c''$$

2. $\int \dfrac{2x+3}{\sqrt{x^2+x+1}}\,dx$

Step 1. To find the values of A and B

$$2x + 3 = A\frac{d}{dx}(x^2+x+1) + B = A(2x+1) + B \qquad \text{-------------- (i)}$$

Equating coeff. of x both sides we get, $2A = 2 \Rightarrow A = 1$

Equating the constants , we get $\quad A + B = 3 \Rightarrow 1 + B = 3 \quad$ ie. $B = 2$

(1) becomes $\quad 2x + 3 = (2x+1) + 2$

$$\int \frac{2x+3}{\sqrt{x^2+x+1}}\,dx = \int \frac{2x+1}{\sqrt{x^2+x+1}}\,dx + 2\int \frac{1}{\sqrt{x^2+x+1}}\,dx = I + 2I' \quad \ldots\ldots (2)$$

where $I = \int \dfrac{2x+1}{\sqrt{x^2+x+1}}\,dx \quad$ and $\quad I' = \int \dfrac{1}{\sqrt{x^2+x+1}}\,dx$

Step 2 To evaluate the integrals I and I$'$

$$I = \int \frac{2x+1}{\sqrt{x^2+x+1}}\,dx = 2\sqrt{x^2+x+1} + c'$$

$$I' = \int \frac{1}{\sqrt{x^2+x+1}}\,dx$$

$$x^2 + x + 1 = x^2 + 2.\frac{1}{2}x + (\tfrac{1}{2})^2 - (\tfrac{1}{2})^2 + 1$$

$$= (x+\tfrac{1}{2})^2 - \tfrac{1}{4} + 1 = (x+\tfrac{1}{2})^2 + (\tfrac{\sqrt{3}}{2})^2$$

$$I' = \int \frac{1}{\sqrt{x^2+x+1}}\,dx = \int \frac{1}{\sqrt{(x+\frac{1}{2})^2 + (\frac{\sqrt{3}}{2})^2}}\,dx$$

$$= \sinh^{-1}[\tfrac{2}{\sqrt{3}}(x+\tfrac{1}{2})] + c'' = \sinh^{-1}(\tfrac{2x+1}{\sqrt{3}}) + c''$$

Step 3 Substituting the values of I and I' in (2), we get

Self Evaluation Test 5

I. Match the following

Column I	Column II
1. $\int \frac{1}{5x+1} dx$	i) $\log(1 - \cos x) + c$
2. $\int \csc x \cot x \, dx$	ii) $\frac{1}{2a} \log \left(\frac{a+x}{a-x}\right) + c$
3. $\int \sqrt{1 + \sin 2x} \, dx$	iii) $\frac{1}{\sqrt{5}} \sin^{-1} \sqrt{5} \, x + c$
4. $\int \frac{1}{\sqrt{x}} dx$	iv) $\frac{a^x}{\log a} + c$
5. $\int \frac{1}{\sqrt{1-5x^2}} dx$	v) $\frac{\tan^6 x}{6} + c$
6. $\int 9^x \, dx$	vi) $\frac{1}{5} \log (5x + 1) + c$
7. $\int \frac{1}{\csc 3x} dx$	vii) $- \log(\csc x + \cot x) + c$
8. $\int \frac{\sin x}{1 - \cos x} dx$	viii) $\log[f(x)] + c$
9. $\int \sec^2 x \tan^5 x \, dx$	ix) $\sinh^{-1} \frac{x}{a} + c,$
10. $\int \csc x \, dx$	$\log (x + \sqrt{x^2 + a^2}) + c'$
11. $\int \cot^2 2x \, dx$	x) $- \cos x + \sin x + c$
12. $\int \frac{1}{\sqrt{x^2 - a^2}} dx$	xi) $\cosh^{-1} \frac{x}{a} + c,$
13. $\int \frac{1}{\sqrt{a^2 - x^2}} dx$	$\log (x + \sqrt{x^2 - a^2}) + c'$
14. $\int \frac{1}{a^2 - x^2} dx$	xii) $\sin^{-1} \frac{x}{a} + c$
15. $\int \frac{1}{\sqrt{x^2 + a^2}} dx$	xiii) $- \frac{1}{2} \cot 2x - x + c$
16. $\int \frac{4x + 6}{\sqrt{2x^2 + 6x + 5}} dx$	xiv) $2\sqrt{2x^2 + 6x + 5} + c$
	xv) $2x^{\frac{1}{2}} + c$

17. $\int \frac{f'(x)}{f(x)} dx$	xvi) $-\operatorname{cosec} x + c$
	xvii) $\tan x + c$
18. $\int \frac{1}{x^2+a^2} dx$	xviii) $\frac{1}{a} \tan^{-1} \frac{x}{a} + c$
19. $\int \sec^2 x \, dx$	xix) $\frac{9^x}{\log 9} + c$
20. $\int a^x \, dx$	xx) $-\frac{\cos 3x}{3} + c$

Write the answers here.

1. …..   2. …..   3. …..   4. …..   5. …..   6. …..

7. …..   8. …..   9. …..   10. …..   11. …..   12. …..

13. …..   14. …..   15. …..   16. …..   17. …..   18. …..

19. …..   20. …..

Answers :

1. vi   2. xvi   3. x   4. xv   5. iii   6. xix

7. xx   8. i   9. v   10. vii   11. xiii   12. xi

13. xii   14. ii   15. ix   16. xiv   17. Viii   18. xviii

19. xvii   20. iv

II. Fill in the blanks

1. $\int x^n dx = \frac{x^{n+1}}{n+1}$ is true when n $\neq$ ……   2. $\int dx = $ …………….

3. $\int \frac{1}{4x} dx = $ …………..   4. $\int 3^x dx = $ …………..

5. $\int \operatorname{cosec}^2 x \, dx = $ …………..   6. $\int \frac{1}{\cos^2 x} dx = $ …………..

7. $\int \frac{\cos x}{\sin^2 x} dx = $ …………..   8. $\int \frac{1}{1+(ax)^2} dx = $ …………..

9. $\int \frac{\sin x}{1+\cos x} dx = $ …………..   10. $\int \frac{\sec^2 3x}{\sqrt{\tan 3x}} dx = $ …………..

11. $\int f'(x) [f(x)]^n dx = $ ………….. where n $\neq$ ………

12. $\int \cot x \, dx = $ …………..   13. $\int \sin x \cos x \, dx = $ …………..

14. $\int \frac{x^8}{1+x^9} dx$ = ...

15. $\int \frac{1}{\sqrt{x^2+a^2}} dx$ = ...

16. $\int \frac{1}{\sqrt{9-x^2}} dx$ = ...

17. $\int \frac{1}{a^2-x^2} dx$ = ...

18. $\int \frac{1}{x^2+a^2} dx$ = ...

19. $\int \frac{f'(x)}{\sqrt{f(x)}} dx$ = ...

20. $\int \frac{1}{\sqrt{1-(ux)^2}} dx$ = ...

Answers

1. $n \neq -1$ 2. $x + c$ 3. $\frac{1}{4} \log x + c$ 4. $\frac{3^x}{\log 3} + c$

5. $-\cot x + c$ 6. $\tan x + c$ 7. $-\csc x + c$ 8. $\frac{1}{a} \tan^{-1}(ax) + c$

9. $-\log(1 + \cos x) + c$ 10. $\frac{2}{3} \sqrt{\tan 3x} + c$ 11. $\frac{[f(x)]^{n+1}}{n+1} + c, n \neq -1$

12. $\log(\sin x) + c$ 13. $\frac{-\cos 2x}{4} + c$ 14. $\frac{1}{9} \log (1 + x^9) + c$

15. $\sinh^{-1} \frac{x}{a} + c, \log (x + \sqrt{x^2 + a^2}) + c'$ 16. $\sin^{-1} \frac{x}{3} + c$

17. $\frac{1}{2a} \log (\frac{a+x}{a-x}) + c$ 18. $\frac{1}{a} \tan^{-1} \frac{x}{a} + c$ 19. $2 \sqrt{f(x)} + c$ 20. $\frac{1}{a} \sin^{-1}(ax) + c$

III. Correct the following errors if any :

1. $\int \frac{1}{ax+b} dx = \frac{1}{a} \log(ax + b) + c$

2. $\int \csc^2(ax + b) dx = \cot(ax + b) + c$

3. $\int \csc(5 - x) \cot(5 - x) dx = -\csc (5 - x) + c$

4. $\int \frac{1}{\sqrt{1-4x^2}} dx = \frac{1}{2} \sin^{-1} 2x + c$

5. $\int \frac{2x-3}{x^2-3x+8} dx = \log(x^2 - 3x + 8) + c$

6. $\int \frac{\csc^2 x}{\sqrt{\cot x}} dx = -2 \log(\cot x) + c$

7. $\int \frac{(\log x)^7}{x} dx = \frac{(\log x)^8}{8} + c$

8. $\int \frac{f'(x)}{\sqrt{f(x)}} dx = \sqrt{f(x)} + c$

9. $\int \frac{\sin \sqrt{x}}{\sqrt{x}} dx = -2 \cos \sqrt{x} + c$

10. $\int \sec 7x \, dx = \frac{1}{7} \log(\sec 7x + \tan 7x) + c$

11. $\int \sin 3x \cos 3x \, dx = \frac{-\cos 6x}{12} + c$

12. $\int \frac{e^{\cot x}}{\sin^2 x} dx = -e^{\tan x} + c$

13. $\int \frac{1}{\sqrt{x^2 - a^2}} dx = \cosh^{-1} \frac{x}{a} + c$, $\log(x + \sqrt{x^2 + a^2}) + c'$

14. $\int \frac{1}{a^2 - x^2} dx = \frac{1}{2a} \log\left(\frac{a+x}{a-x}\right) + c$

15. $\int \frac{1}{\sqrt{x^2 - 64}} dx = \cosh^{-1} \frac{x}{64} + c$, $\log(x + \sqrt{x^2 - 64}) + c'$

Answers

1, 4, 5, 7, 9, 10, 11, 14 are correct.

2. $-\frac{1}{a} \cot(ax + b) + c$

3. $\cosec(5 - x) + c$

6. $-2\sqrt{\cot x} + c$

8. $2\sqrt{f(x)} + c$

12. $-e^{\cot x} + c$

13. $\log(x + \sqrt{x^2 - a^2}) + c'$

15. $\cosh^{-1} \frac{x}{8} + c$

www.ingramcontent.com/pod-product-compliance
Lightning Source LLC
Chambersburg PA
CBHW081208180526
45170CB00006B/2253